The DVD Novel

The DVD Novel

How the Way We Watch Television Changed the Television We Watch

Greg Metcalf

PRAEGER

AN IMPRINT OF ABC-CLIO, LLC
Santa Barbara, California • Denver, Colorado • Oxford, England

Library of Congress Cataloging-in-Publication Data

Metcalf, Greg.
The DVD novel : how the way we watch television changed the television we watch / Greg Metcalf.
p. cm.
Includes bibliographical references and index.
ISBN 978-0-313-38581-0 (hardcopy : alk. paper) — ISBN 978-0-313-38582-7 (ebook : alk. paper) 1. Television series—United States—History and criticism. 2. Television viewers—United States—Attitudes. 3. DVD-Video discs—Social aspects. I. Title.
PN1992.8.S4M395 2012
791.45'6—dc23 2012006423

ISBN: 978-0-313-38581-0
EISBN: 978-0-313-38582-7

16 15 14 13 12 1 2 3 4 5

This book is also available on the World Wide Web as an eBook.
Visit www.abc-clio.com for details.

Praeger
An Imprint of ABC-CLIO, LLC

ABC-CLIO, LLC
130 Cremona Drive, P.O. Box 1911
Santa Barbara, California 93116-1911

This book is printed on acid-free paper ∞
Manufactured in the United States of America

DEDICATED TO Frank Black, James Caldwell, John Caravella, Martin Castillo, Anthony Clifton, Richard Cross, Reginald Cousins, Silvio "sadness accrues" Dante, Arthur Dent, Lester Freeman, Lorelai Gilmore, Jeremy Goodwin, Latka Gravas, Trevor Hale, Gregory House, Omar Little, Philip Marlow, John Munch, Arthur Parker, Frank Pembelton, Reginald Perrin, Jim Profit, Dunstan Ramsay, James Rockford, Alexander Scott, Kalinda Sharma, Jane Tennison, Geoffrey Tennant, Walter White, and Lionel Whitney.

And, always, Adena Watson.

Be amusing: never tell unkind stories;
above all, never tell long ones.

—Benjamin Disraeli

Contents

Introduction — Because "Excuses" Sounds Too Defensive a Way to Start, ix

Chapter One — Television Is an Object and a Narrative Form, 1

Chapter Two — *The Singing Detective* and British Television, 15

Chapter Three — Steven Bochco and Too Many Stories, Not Enough Time, 37

Chapter Four — Stealing the Soap and Longform Network Television, 63

Chapter Five — Premium Blend Genre and Longform Network Television, 81

Chapter Six — Men Behaving Badly on Cable, 97

Chapter Seven — Comedy on Television, 121

Chapter Eight — Longform Comedy on Television, 135

Chapter Nine — Graphic Novels and DVD Novels, 155

Chapter Ten — David Milch and Television as Literature, 167

Chapter Eleven — David Simon and Fictional Nonfiction Television, 187

Chapter Twelve — Endings, 205

Notes, 215

Bibliography, 221

Index, 223

Introduction

Because "Excuses" Sounds Too Defensive a Way to Start

We aren't watching television anymore.

I don't mean that in the standard, "viewership is down due to videogames and work schedules" sense.

I mean two other things.

The shows we watch on television now are fundamentally different from what television was in its Golden and even its Silver ages.

At the simplest level, American prime-time television through the 1970s meant freestanding episodes with characters that never changed. Now almost every narrative show on television—cable or network—has at least one long story arc for the series, and possibly more within a season.

A 13-hour story allows for a different sort of story and storytelling than presenting 13 one-hour stories. Episodes become chapters. Characters can be developed and transformed in the way we expect characters to change in traditional literature and drama. Themes and counterthemes can be juxtaposed and explored on a much larger landscape that allows for contradiction and ambiguity which would be out of place in traditional television writing.

But I also mean that we aren't watching television.

We're watching online streaming video and TiVo, and we are watching DVD sets. These shifts in technology, viewing habits, and storytelling have influenced one another to create a new form of television that has emerged on television but is something different from "television." When viewers can watch as much of a series as they want rather than measuring it out in weekly episodes, writers can treat a season of television as a single story.

It's no longer worthy of comment that people are only now starting to watch *The Sopranos*, years after it ended, or that my friends go to the beach and the husband finds *Treme* on iTunes and watches the whole first season that night, then watches it again the next day. Or that Mark Maron tells a guest on his podcast about watching an hour or two of *Six Feet Under* every night before he goes to bed, or that George R. R. Martin (writer of *Game of Thrones*) hedges his criticism of the ending of *Lost* by explaining that he hasn't watched it on DVD yet but only saw it when it was broadcast on NBC.

It's not even surprising that, when AMC decided to promote the fourth season of *Breaking Bad* during the finale of its series *The Killing,* they ran a commercial in which Meillos Enos (star of *The Killing*) admitted that she hadn't watched the first three seasons of *Breaking Bad* until she watched it all on DVD.

It's worth thinking about that for a moment.

American television makes its money by charging for advertising based on how many people watch the shows when it is broadcast. AMC's ad for *Breaking Bad* endorses watching the series on DVD. With no commercials.

The latest DVD sets of *Southland* and *Breaking Bad* are promoted as uncensored. That means what you see on television isn't the "real" version of the series, you only get that on the DVD set. Everything from the mysterious drama of *Lost* to the wacky comedy of *Community* is sold as being better on a DVD set than it is on television.

These networks are no longer promoting television shows, they are promoting video narratives that are aired on television before the real version is made available on a DVD set. In fact, *Fringe* was allegedly canceled in the spring of 2011 only to be un-canceled and renewed after the network reconsidered the number of viewers watching the show on TiVo and DVD.

Television used to be ephemeral. An episode aired and, if you missed it, you might have one more shot to see it during a summer rerun. If you were lucky, the show might be sold into syndication and with constant vigilance you might see the episode in rerun, but that was your only chance.

But now television seasons exist forever. Not the episodes, the seasons. The final state of a television show is a boxed DVD season set, about three hours to a disc, with no commercials to interrupt or pop-up IDs and promos to obscure the screen. You can pause it or watch it all in one sitting, if you get distracted and miss something, you can rewind and watch it again.

The true state of a television series is, paradoxically, no longer on television. It is a set of DVDs that sits on your shelf in a box that looks like a book. And increasingly that's how people watch them.

Watching these shows has become like reading a novel.

The people who make television shows know the end of their work is a DVD set. Writing a story for individual episodes that air once, a week apart, and are gone forever is a different sort of storytelling than creating a 12-hour story that can be watched in extended sittings and rewatched as often as someone likes. It's the difference between a collection of 12 short stories and a novel that is 12 chapters long.

This does not mean these changes are embraced by everyone making television or that these longform shows are inherently artistically superior to episode-based television. There are bad novels and great short stories. The point is that American television is increasingly writing novels.

In the summer of 2011, television reviewers and commentators confirmed the change in a cluster of negative season-end reviews and comments about *Game of Thrones, Treme,* and *The Killing*. One hero was too slow or missed clues that a "television detective" wouldn't, there was too much gratuitous nudity, and the wrong characters were killed, or the characters didn't fight back fast enough and it was hard to remember what happened a couple weeks earlier. The reasons varied but the underlying theme was that these shows were bad television or were not "television" because they did not follow the (unwritten) rules of television. Instead they did things that were very common in detective, fantasy, or literary novels or plays or even some movies but were not common on American network television.

The complainers were right. These shows are not what Americans have always meant by "television." They are something else.

This book tries to explain what that something else is and how American television got to a point where it makes shows that aren't television.

If you want the summary of this book in one paragraph, it's this: At least as far back as the 1980s, people who make American television are feeling the limits of the form and trying to make it something else. As they push the bounds of what television is, the model that continues to arise is to write the show as a novel, with episodes as chapters. The struggle to create a sort of storytelling that isn't "television" but for practical reasons has to be made for television becomes easier with the arrival of the DVD set. The result has been a change in the way that almost all television fiction is written but also the development of new strategies of complexity and depth and literary qualities that were not previously available on American television.

David Simon articulated the tension between the necessary weekly episodes and the single season story in answer to complaints that *Treme* was not producing satisfying individual episodes by saying that he has to "accept that you're writing the show for people who have a complete season DVD set in front of them, or who are watching the show via HBO On Demand, or who can otherwise absorb it all as a piece, and watch [the episodes] all in a row."[1]

That is the DVD Novel.

That Which Follows

"DVD novel" is a lousy term.

But all the others are worse.

The "digital novel" sounds like something on a Kindle, which, admittedly, may be where we end up watching television before long. The "television novel" or the "video novel" sounds misleadingly retro. "Nonprint novel" would have the advantage of alliteration, but it doesn't acknowledge the important change that the DVD set of a television series makes in how we understand television.

The term probably comes from the graphic novel. Just as a graphic novel is the compilation of a series of comic book episodes bound in one cover to enhance their single story, the DVD novel is the compilation of a season of television episodes into a DVD set that stresses their unity as a single long, coherent story.

What follows is a consideration of how American television, which had been focused on creating series made of ephemeral freestanding episodes that could run forever without change, began to tell different sorts of stories with multi-episode, longform stories with change and a definite ending. The examples are somewhat arbitrary and were chosen because they illustrate changes, not necessarily successes. There is no claim that these are the best shows of the last few decades—although they tend to be well-written shows, only that they are part of a movement toward telling different types of television stories.

Chapter 1 lays out a blended history of the interrelated development of American television storytelling, economics, and technology. The popular tradition of turning episodic stories into longform narrative is considered as well as the essential conventions of soap opera.

Chapter 2 points to differences between British and American television by way of setting a context for Dennis Potter's *The Singing Detective*, a seven-hour show that had a significant impact on television makers and viewers when it aired in the United States in 1987. The influences of theater, social documentary, novels, and the short series are considered as creating very different expectations of television narrative in Britain than in the United States. The significance of *The Singing Detective* and evidence of its influence on American television are discussed.

Chapter 3 considers shows created by Steven Bochco, a champion of the work of Dennis Potter, who is credited with reinvigorating genre drama on American television with his advancement of longform television by incorporating soap opera elements into the cop show *Hill Street Blues*, which became the model for reviving genre television. Bochco's contributions to the creation of dramedy, his musical cop show, *Cop Rock*, in the context of music on American television, and his full-season, single-case drama *Murder One* are examined as failures of television that gave permission for more recent programming such as *Glee* and *Damages*.

Chapters 4 and 5 consider longform network television. Chapter 4 examines the transformation of the anthology show into a longform narrative, from the basics of *The Fugitive* in the 1960s to the Mythos of *The X-Files* in the 1990s. Blending genres is considered as a tool and consequence of longform storytelling. *Wiseguy* (1987) is discussed as a breakthrough in the establishment of television antiheroes made possible through the character development afforded in multi-episode storylines. The show's sympathetic villains laid the groundwork for Tony Soprano, Dexter, and Walter White of *Breaking Bad*.

In Chapter 5, *Veronica Mars, Gilmore Girls,* and *House* are considered as different approaches to blending genres to create series that satisfy in weekly viewing ("the short story") while also creating "chapters" in a season-long story that reveals itself, and changes the story, when watched on DVD as a novel.

Chapter 6 considers the HBO hits *Oz* and *The Sopranos* in the context of male-centered soap operas and episodic versus longform narrative as it develops on cable television, premium cable content disguising the way that the new shows continue to use network television structures. Comparison of *The Sopranos, The Shield,* and *Breaking Bad* is made in terms of the aspiration toward Shakespearean (and Chekovian) television and the idea of television tragedy in the postheroic age. The difference between watching a single episode and immersion in *The Sopranos* is explored. The structural role of humor in *The Sopranos*, its absence from *The Shield*, and its gradual removal from *Breaking Bad* are also considered. The chapter concludes with a discussion of the strategy of not revealing feelings in soapish shows such as *Mad Men* and *The Good Wife*.

Chapters 7 and 8 discuss television comedy. Chapter 7 considers the conflicts between situation comedy and longform narrative. The birth of one form of dramedy in the syndicated viewing of sitcoms is explained. The overtly soap opera sitcoms *Mary Hartman, Mary Hartman* (1976–1977), and *Soap* (1977–1981) are considered as stages on the way to *The Days and Nights of Molly Dodd* (1987–1991), the series that set the model for the attitudinal comedies of cable television.

In Chapter 8, after considering examples of the novelistic model of television comedy in England and Canada, the importance of the ironic attitude of the protagonist rather than the situation in longform television comedy is explored, where attitude replaces jokes. The gendering of longform comedy is considered by juxtaposing shows such as *The Big C* and *Nurse Jackie* with *Entourage* and *Californication*. Reality comedy is examined in the British version of *The Office* and *Curb Your Enthusiasm*.

Chapter 9 considers other longform television attempts based in the logic of the graphic novel. Comics-based television plotting, such as *Heroes* and *Lost*, is discussed before moving on to television series using the pacing and visually-based storytelling of graphic novels in *Carnivale, Rubicon,* and *The Killing*. Graphic novel adaptations are discussed, as well as the model of online comics and webisode series. The strategy of continuing canceled series as graphic novels concludes the chapter.

Chapter 10 considers shows by David Milch, in the context of his academic background and his literary aspirations. Milch is viewed as an author whose interest is increasingly in the transcendent, rejecting the traditional individual hero of television while trying to write novelistically in the corporate medium of television. The rejection of television tradition in the language and community of *Deadwood* leads to *John from Cincinnati*'s philosophical exploration of some of William James's themes in the form of the magical realist novel.

Chapter 11 considers David Simon from his journalistic background as a writer who develops a form of reporting through fiction, a "New-er" Journalism that documents the collapse of the institutions of the modern city. Simon's early declaration that *The Wire* would be a "video novel," as he recruited novelists to help him write the show, makes him the clearest in his rejection of network television storytelling. *The Wire* redefines heroism in a series where criminals and cops share the same frustrations within a depiction of the modern city in which network television roles and rules do not apply. *Treme* is seen to expand on those ideas to tell the stories of the average residents as it champions the community and small act of recognition over the artificial drama of television. *Treme* most explicitly rejects the idea of the single episode to tell a multi-season story of the gradual rebuilding of New Orleans's communities after Hurricane Katrina. In the process, it attempts to retrain viewers to watch television the way they might read, to expand their expectation of the rewards of the DVD novel.

Chapter 12 concludes with a consideration of the continuing move toward longform storytelling in both television and film and what it means that television and movies end up sitting next to each other on a DVD shelf. The American trend of adapting novels to television is considered as a triumph of the influence of the DVD set as a novel, with emphasis on examples from Stephen King, George R. R. Martin, and Neil Gaiman. The possibilities of direct-to-DVD and longform Internet television are briefly considered on the way to an anecdote on the possible legacy of *The Singing Detective*.

Chapter One

Television Is an Object and a Narrative Form

In Which We Consider the Development of American Television, Serial Narrative, and the Soap Opera

Sex on television can't hurt you unless you fall off.

—Oft-repeated joke of the second half of the 20th century

Television is an object through which audiovisual narratives are delivered.

It was originally a piece of furniture (a box), became more of an appliance, and is now supposed to be hung on the wall in imitation of a movie screen. Or a painting, if you prefer.

Televisions have started to act increasingly like something other than TVs. The screen size has long since changed to proportions of the widescreen of movies, which, ironically, went wide in the 1950s to distinguish film from television.

The last television I bought was cheap. It only had four HDMI inputs, two video inputs, two component inputs, and one PC link. It is physically the screen for my computer, DVD player, VHS player, online access, and video games.

Where television used to be broadcast free in America, I now can't actually watch "television" on the television unless I pay more money to a cable company that happens to own NBC as just one of its media assets.

Which brings us to the second meaning of television.

Television is a type of storytelling that developed to serve the commercial delivery systems that grew up to provide content to be viewed through that piece of furniture.

In the United States that storytelling settled into hour or half-hour blocks of narrative interrupted by blocks of commercials that had nothing to do with the content of the shows, except that they were the reason the shows were there. The shows were created to deliver the eyeballs of certain consumer demographics to those commercials.

As film fought to hold on to an audience in the face of television, they didn't just widen screens and try stunts like 3-D movies to get the audiences to leave their homes and go to the theaters. Film's narratives also came to emphasize the same sense of spectacle; big pictures, big dramas, and big stars.

By contrast, television was small and its stories reinforced a domesticated vision. Situation comedies, most obviously, delivered the idealization of the living room culture of their audience. Television stories gave you predictability and calm in contrast to the excitement and danger of films.

Certain genres emerged—the soap opera, the cop show, the legal drama, the mystery, the situation comedy—and the programs tended to reinforce the dominant values of the culture. "Father Knows Best," the police are your friends, justice triumphs. The "messages" of television reinforced the beliefs of the audience, made them feel comfortable and safe.

In the service of this function, every episode of a television series was supposed to stand alone. Television was ephemeral, of the moment of its broadcast, then gone. You could not rely on viewers to watch every episode or to remember information from week to week, so storylines had to conclude in a single viewing. Everything had to be tied up and put back in the box. This created a dramatic limitation on writing. The main characters had to remain the same from episode to episode. They had to always be the same so viewers saw familiar people acting the same whenever they came back.

This is the antithesis of traditional dramatic art and literature, where the audience watches the lead character(s) transformed by a series of interactions. Television series demanded that there be no transformation, only constancy. It requires a special skill to create stories without change and that became the art of television writing.

This is a grossly oversimplified explanation of the history of American television up to the period where this book begins.

The 1970s brought widespread use of the television remote control, allowing lazy viewers to channel-surf rather than being trapped by inertia (physical or narrative) with the "least objectionable" program. Hard as it may be to conceive today, viewers had to stand up, walk to the box, and physically turn a dial every time they wanted to change to another channel. The remote control altered the truism of programming that viewers would pick a channel and stay with it for the night; they could now change channels any moment they got bored.

The remote control eventually required a change in commercial television writing that became clear in the 1980s. With viewers able to effortlessly change channels the moment they became bored, or tune away during commercials and never come back, the way a story was told changed.

Previously, a television show had to hook its audience quickly with a dramatic opening, but had latitude with the rest of the show, assuming that the audience would stay once hooked. The remote control drove the focus to writing for the segment between commercials. The writer needed to hook the audience after each commercial, offer some sort of dramatic arc that ended with enough suspense that the viewers would stay through the commercials, or at least come back after they were over.

The 1970s also brought premium cable television. HBO launched in 1971, the Star Channel/Movie Channel in 1973, Showtime in 1976, and HBO's Cinemax in 1980.

As the names implied, these were channels offering people the opportunity to watch recent feature films in their home for a fee. (HBO stands for Home Box Office.) HBO went to a 24-hour programming day at the end of 1981 and made its first original programming in 1983. (*The Terry Fox Story* was their first movie and *Fraggle Rock* their first episodic show.) Premium cable also turned the weakness of using reruns to fill out a programming schedule into the strength of the greater availability of programs to viewers who couldn't watch a show broadcast only once.

Even when premium cable was mainly airing recently released feature films, the existence of pay cable forced network television to adjust to hold onto their audiences.

The networks' business model had been based on "broad casting" sending the same stories to everyone and making money on the commercials shown to those masses. Premium cable "narrow cast" their programming to a select group that paid for their access. In response, even as they continued to "broad cast," networks began to emphasize capturing the *right* eyeballs, targeting the most prized demographic, rather than appealing to everyone. Fox built its original network audience by targeting the demographic groups—nonwhites, urbanites, and young adults—that the established networks had neglected.

The networks started allowing more "adult content"—stronger language and adult situations—to combat the more limited censorship on pay cable stations but, in the process, they opened the door to writing television in a different way. Television writing was now being asked to directly compete with film writing, with a special concern with holding on to adult/"sophisticated" viewers. (Ironically, as it became apparent that a television airing was part of a popular film's life and profitability, films began to be written with the same emphasis on the short segment between commercials that they would eventually inhabit. *Armageddon* (1998) may be the purest example of this short attention span film writing.)

One response by the networks was the miniseries. This was the American name for a phenomenon that had been central to British programming since the 1960s, the adaptation of a novel into a multi-episode television series that was shorter than a normal series, with a bigger per-episode budget, that told a story with full dramatic structure and closure in contrast to the continuing stories of most American television. Examples include *QB VII* (1974), *Rich Man, Poor Man* (1976), *Roots* (1977), *Shogun* (1980), and *Winds of War* (1983). The miniseries were "event television" that acted more like movies and were based on novels.

Cable and network began to bring in film directors, writers, and actors to provide film quality that set their television shows apart from other television shows.

The 1970s brought the video recorders, video home system (VHS), and Betamax. By the 1980s, the video cassette recorder (VCR) was entrenched in American homes and changing the function of the piece of furniture and changing the narratives that could be seen on "the box." Viewers could also record a show and then fast-forward through commercials and watch just the content they wanted, when they wanted. Or they could ignore television and watch rented movies on their television box without paying HBO or leaving their house.

Or they could have friends with HBO record *The Sopranos*. This was a significant distribution pattern of the shows on premium cable. It also led to people viewing more than one episode at a time, as a VHS tape could hold two to six hours of content, depending on recording speed, and people who viewed HBO in this way were always catching up with the paying viewers of the shows.

HBO picked up on this practice and began releasing *The Sopranos* series in VHS sets in 2000, increasing profits from the show and encouraging the transition from thinking of the initial airing as the true version of a show to the show's recorded set as the authoritative version of a television show. There had been other VHS season sets, for example ABC's *Twin Peaks* was released 10 years earlier, but it was rare. Most commercial videos of television were "greatest episodes" collections of programs such as *The Honeymooners, I Love Lucy,* or *Mary Tyler Moore.*

The release of the entire season of a television show on VHS, in a high-quality recording, suggested that this television series was not an ephemeral entertainment. It was to be a part of the viewer's home library, and implied that the viewer would watch, and rewatch, the program in a manner closer to the reading and rereading of a favorite book.

This is a fundamental shift in how we understand a television program, the idea that it has permanence, and that it is worthy of permanence.

Videotape also made it easier for everyone, including television writers and producers, to see programs from other countries that made television differently, and films that didn't make it to the local theaters. True, there was a need for special video players that could handle the PAL-encoded videotapes that were produced in most countries outside of the United States (which uses NTSC-encoding), but it was now possible for Americans to see British television that might not be run on their local PBS (Public Broadcasting Service) station.

It is in this period that what we call television begins to mutate into something else, a transformation that continues to take place as the meaning of television is in a period of redefinition, if not collapse.

The technology has continued to change what we consider television.

By the end of the 20th century, the pervasiveness of the Internet in the digital fin-de-siècle means that no one needs a television (the box) to watch television. Television programs are just one more sort of content streaming through the tubes of the Internet to be downloaded or streamed at any time and, increasingly, any location. My cable company tells me that I need to stay with them because I can watch television, movies, music, video, and sports on my WiFi-ed laptop or my telephone, wherever I happen to be, raising the question of whether it's a television show at all anymore if it was produced for the Internet and watched on a telephone.

TiVo, the digital television recorder, was introduced in 1999 and surpassed the VCR in ease of recording programming and the capacity it could hold without taking up any additional space in one's home. If a viewer wants to record beyond the large capacity, they can simply add an external hard drive. The viewer can program various recording combinations from whatever their cable provider offers but a significant, and regularly used, feature allows a viewer to easily designate a television show and have every episode reliably recorded for future viewing. One result has been the practice of recording entire seasons of a series, to be watched in extended viewing sessions after the season has ended.

By the end of the 20th century, the DVD had largely replaced the videotape. (The Blu-ray disc is considered a subgroup of the DVD for the purposes of this overview, even though it was intended as a replacement for the DVD.) The DVD was so inexpensive that it became cheaper to buy a disk than to attend a film in a theater, with the increasing likelihood that an American would have their own DVD "library" of their favorite films or television shows. The easy availability of films and television series on DVD for cheap purchase or mailed rentals through Netflix or Blockbuster or streaming movie rentals online to your television destroyed video stores. It also threatened the initial appeal of the premium cable channels. While they continued to air feature films, HBO and Showtime highlighted their expanded creation of original programming.

By 2005, the release of the boxed DVD set of a television season had become a significant part of the business model and programming schedule of both cable and network television programs. While *24* successfully released a season on DVD the day after the last episode was broadcast, the convention has become that the previous season's DVD set of a hit show is released shortly before the new season begins to air on television, with the advertising for the previous season's DVD release cross-promoting the new season. But "failed" series, such as *John from Cincinnati* or *Firefly*, will also receive boxed sets through which they can develop a longer and more successful afterlife.

In the latter case, the DVD set gave fans of *Firefly* their first chance to see all of the episodes that had been filmed and to see them in their intended order. (Fox had shuffled the order and canceled the series with unaired shows.)

There are several important shifts in thinking here. The importance of the DVD set is that it formalizes the idea that what we thought of as television—the ephemeral programs broadcast onto the box in the living room at one time on one evening (unless the show reran or went into syndication) is not the true state of television shows. The version that appears on the DVDs, in the order they appear on the DVDs, is the true version of the program.

The most insignificant difference is perhaps the most telling. I can always tell I am watching television because there's at least one irritating bug—a station logo, a network logo—on the screen. On some channels, the bug is joined by pop-up commercials and promotional graphics that can cover as much as a third of the screen. The visual content of a television program as it was shot cannot be seen as it is aired; it can only be seen correctly on a DVD set.

Perhaps more important, since the turn of the century, creators of television shows have known before they began that the program would end up on a DVD set, with several episodes to a disk, boxed together in packaging that makes the season of the show look like a book on a shelf.

The broadcast of a program is only a transitional state; the final state of these programs is the DVD set.

Increasingly, cable and network television creators have been in the business of producing longform narratives that will first air as a weekly television program, but their lasting impact will come when they are assembled into a season-long work that can be watched more intensively on DVD sets. Premium cable series provide obvious examples of longform narratives that reward more intensive viewing on DVD; however, networks have developed shows such as *House* and *The Good Wife* that initially function as episodic entertainment but reveal new depths and dramatic arcs when watched on DVD without commercial and broadcast interruptions.

The form of the DVD set shapes the story.

The case, like a trade paperback book, is taller than wide in a way that does not reflect the circular shape of a DVD but instead invokes all the cultural signification of a book. Each "page" has one or a couple DVDs, which leads to another way that the form changes the meaning.

Patterns of DVD consumption, especially movies on DVD, train viewers toward an assumption of "serving size." One DVD is a viewing experience. With a film, that means 90 minutes to two hours as a general frame of reference.

Two or three "hour-long" episodes are commonly put on a DVD in a DVD set. If a network show, that means that two episodes of the series are about 80 minutes, shorter than the shortest feature films. Three episodes are about two hours long, an appropriate movie length. The physical format guides the viewer toward watching more than one episode at a sitting.

But the form also makes the important transition of turning the episodes into chapters in the "book" of the DVD set, guiding the viewer to experience the episodes in a novelistic experience of reading chapters that build toward an ending at the end of the "book" of the DVD set, a story that ends at the end of the season.

Episodes of series become chapters of a season rather than the standalone short stories of the past.

Serial Art Compiled

The reassembly of short episodes into a single larger work is not unique to the DVD novel. In fact, it is a common stage that has been repeated across various popular media to create a more substantial—and profitable—work of art. Leslie Fiedler claimed that the novel is the "first art form that is an honest-to-god commodity."[1] Clearly serial novels provide an example of this commercial art.

The Victorian novels of Charles Dickens's day were first published in ephemeral monthly installments. The standard 32-page "episodes" included commercials—ads inside the front and back covers—although the final volumes did not. There were problems of pirated issues where authorized publication didn't satisfy demand, just as today episodes of the BBC's *Life's Too Short* (2011–) and HBO's *Boardwalk Empire* (2010–) are posted on unauthorized streaming video sites as soon as they are "published" in each week's broadcast.

When the novels were completed, they would be republished in more expensive editions of one to three volumes (by way of comparison, *The Wire* is a five-volume collection). Just as with DVD sets today, the "boxed set" novels of Dickens had added features—a hard cover, decorative edging on the pages, and extra illustrations. Dickens and his contemporaries would also release "director's cuts" as they made corrections with the new editions.[2]

This system allowed publishers to maximize profit by, essentially, selling the same content twice, the second time in a more permanent upmarket form. The limits of the publishing world meant a thriving business in bootleg copies of popular novels that parallels the selling of unauthorized DVD sets and online downloads of premium cable shows.

Serial storytelling didn't end with *Martin Chuzzlewit* and *David Copperfield.* In the next century, hard-boiled detective fiction was developed in short stories printed in cheap pulp magazines like *Black Mask.* Dashiell Hammett and Raymond Chandler would reassemble their individual stories into longform narratives, making minor adjustments to string the stories into novels such as *Red Harvest, The Maltese Falcon,* and *The Big Sleep.*

The ephemeral episodes became something else when put between covers, a single work that allowed for far deeper reading in context, without interruption. The extreme example of maximized profits from reselling is found in the form of the American comic book, initially created by a newspaper publisher reprinting the comic strips from the paper. As comic strip collections became comic books, they also created demand for original stories, rather than recycled strips. As the stories developed into chapters of a larger story, the comics themselves were reassembled into graphic novels, making them more permanent and maximizing the potential profit from a single work of art.

In the same way, the DVD set as a novel, the reassemblage of a season of a television series, starts to show new ways to write—and to watch—television shows. The semantics of the television show versus the DVD collection is essentially the same one that the comic book world has already gone through. For the premium comic books, the release as individual issues is only a transitional moment on its way to the permanent, and the "more legitimate" literary form of about six issues compiled into a single graphic novel. Comics purists will collect and prize their separate, vinyl-wrapped issues, but the general audience market is in the graphic novels. Public libraries now shelve graphic novels amid the traditional text novels, and graphic novels are regularly reviewed in literary journals and the popular press and taught as literature in universities.

Whether graphic novels are literature is no longer a question, although some might argue about "comics." The difference is the length and packaging. Within a couple hundred pages of comic book storytelling, there is room for character and thematic development, secondary themes, subtext, complexities, resonances, and payoffs to first act Chekovian guns over the fireplace, the sorts of developments and closures that we expect of a novel. The recognition of the expanded narrative possibilities, and the creative permission offered by the commercial success of the graphic novel form, has led to individual comics being seen less as the traditional collection of individual stories and more as chapters in a longform story, a novel. That change in understanding of the narrative unit has broad implications.

In the same way, the recognition that all 12 or 22 episodes of a television show are not individual stories, with the expectations of independence and closure, but are 12 or 22 chapters of a single novel, with each chapter advancing the story, fundamentally changes how a series is written. Most television creators of the moment try to balance the need for a closed experience from each chapter episode with a sense of a longer story being told. David Simon stated this most explicitly when he said that *Treme* was made to be watched in multiple-episode viewings on DVD or on-demand cable, not as a weekly broadcast. Others don't say it, but their DVD sets reveal the longform narrative of the series.

Sound recordings went through the same development. Once the expanded form existed, people found new ways to explore its possibilities. Jack Benny would reissue his radio show routines on 78 rpm recordings in an actual album to create a more permanent object and to create another revenue stream from the same material. Over time, the idea of creating a record album of thematically-related songs, Frank Sinatra's *Songs for Young Lovers* (1954), for example, would lead to the creation of the concept album.

The concept album shifted the assemblage of a dozen unrelated songs into the chance to create a larger experience of 12 related songs. *Pet Sounds* (1966) by the Beach Boys and the Beatles' *Revolver* (1966) were seen by their advocates as the first successful rock concept albums. For some it was the conceptually related songs of *Sgt. Pepper's Lonely Hearts Club Band* (1967), for others, the narrative rock opera of the Who's *Tommy* (1969), or, somewhere in between, *Darkness on the Edge of Town* (1978) from Springsteen.

As John Landau explained about the making of Bruce Springsteen's *Darkness at the Edge of Town*, "The work of art is the album. It's not any one element. It's not a cut. The highest high is the complete album. That is the highest developed thing in rock."[3]

By the 1980s, it was assumed that the songs on an album were conceptually linked, creating a "long arc," to use the DVD novel term. The development of the autobiographical voice in albums as diverse as Tupac Shakur's *Thug Life: Volume 1* (1994) to Taylor Swift's *Fearless* (2006) leads to the reading of the album as a single personal statement. As digital downloads replace CDs as the medium of music, the freestanding song increasingly returns as the unit of popular music, but the idea of the album remains.

The Internet has also created new economic models for seriality. Graphic and traditional novels are published as a page or a chapter at a time on websites to create interest. When the story is completed, they are then published between covers and sold to many in the same online audience and to more traditional book buyers. David Wong's *John Dies at the End* (online starting in 2001, published in 2007) and Catherynne M. Valente's *The Girl Who Circumnavigated Fairyland in a Ship of Her Own Making* (2011) are two traditional novels that followed this model of giving away the content serially before selling it. Graphic novels following the same model include the later *Talisman* books by Carla Speed McNeill and Mark Siegel's *Sailor Twain, or, The Mermaid in the Hudson*.

Soap Opera and Serial Television

The serial narrative also simply expands the possibilities for storytelling. One of the oldest forms of serial storytelling on American television, going back before television to radio, is the soap opera.

Soap operas are serial melodramas aimed at a stay-at-home female audience. The television soap operas (getting their name from early sponsorship by soap companies) were a natural continuation of the radio soap operas and also domesticated the very popular Hollywood genre of women's melodramas from the 1930s through the 1950s when adult women were considered a significant part of the film audience.

These melodramas were also known as "women's weepies," and they served an important cultural phenomenon. They told of a woman unable to fulfill her societal role as a mother, a daughter, or a lover/wife. Women in these stories were forced to sacrifice their own dreams for others, and this suffering was noble.

Two obvious effects of these programs were to reinforce a traditional, Victorian definition of women and the women's sphere and to provide a cathartic release. Furthermore, however bad a viewer's life was, it was better than the surrogates in the soap, so she should be content to continue to live her own life of lesser sacrifice. Given the audience and the message, it is not surprising that the stories focused on the emotional lives of women in women's spaces. The focus was primarily the family—or potential family—and the home. When women left the homes, they might go to hospitals or restaurants, female spaces of nurturing and comfort. Males tended to be love objects, authority figures, and betrayers.

The Shift to Male-Centered Soap Opera

One traditional analysis of the change in prime-time television in the 1980s is that Steven Bochco and his collaborators in creating *Hill Street Blues* (1981–1987) for NBC showed the way to use the continuing storyline structure of the soap opera to re-energize the exhausted nighttime genre stories like the cop show, the legal drama, and the medical drama.

What is often overlooked in this analysis is that this absorption of the soap opera required a sex change. Daytime television was aimed at women. Nighttime genre television was aimed at men.

It wasn't just mixing soap opera and cop show that mattered. The more adventurous daytime dramas had already begun to add cop show elements—crime stories and mysteries—to the soap opera structure.

The innovation of *Hill Street Blues* was to turn the soap opera's female point of view to a male point of view. The world of suffering moved from the Victorian female sphere—the comforting home environment—to the male sphere—the hostile workplace.

With the shift in locale came a shift to the suffering of men as they were unable to fulfill *their* culturally defined role. They suffered from the inability to succeed, to control, to bring order, to win, and to mete out justice, and to find—or be—a suitable mate.

The audience surrogates for these stories are "sensitive men." Readers who lived through the early 1980s may remember Phil Donohue and Alan Alda presented as models of a new sensitive male, an ideal that quickly became a cliché, leading to best-selling backlash such as the *Real Men Don't Eat Quiche* book.

The hero that develops with *Hill Street Blues* is the traditionally tough-guy hero of the cop genre, but one who reveals his emotional side and whom we see suffer. This is the model of almost every television series now, so it may be hard to grasp a time when the heroes of television had no emotional life, only jobs—heroes such as Sergeant Friday of *Dragnet* (1951–1959), Elliot Ness of *The Untouchables* (1959–1963), and Napoleon Solo and Ilya Kuryakin of *The Man from U.N.C.L.E.* (1964–1968). Even the members of the all-male Cartwright family on *Bonanza* (1959–1973) concealed their inner lives to live outer lives of honor and duty.

Today, almost any television series presents men shaped by emotional realities. It might be one of the many detectives tormented by self-loathing and the desire for revenge for the murdered lover they were unable to save, or the love for a coworker they can never express, or a childhood trauma, or an addiction, or guilt from the death of members of their unit when they survived. Or a combination of a couple of them; guilt and inability to express love to a colleague is probably the default network male dramatic character at this point.

This is even true of most comedies. Part of the appeal of *Curb Your Enthusiasm* (2000–) may be that it presents the situations of the "reality" show soap opera—the social discomfort and petty conflict addictions—but at the center is Larry David, a character who heroically rejects introspection or sensitivity.

While the soap opera form has permeated most of prime-time programming, the female-centered daytime dramas continue to lose their audience. Some have explained this as a loss of the stay-at-home daytime audience, but it seems more likely that the emotional function of soaps has been replaced, first by Oprah Winfrey and her less reputable imitators who offered "real" stories of human suffering and triumph, with the focus shifting from the suffering guest to the empathetic host, only to be replaced by so-called "reality" programming and soap-ish game shows.

One of the strangest adaptations of the serial soap opera conventions continues to be seen in the turning of the game show into a longform television narrative. Game shows used to be an episodic distraction, but they have developed into serial melodramas of manipulation, suffering, and emotional betrayal. Whether *Survivor* (2000–) or *The Amazing Race* (2001–), *Project Runway* (2004–) or *American Chopper* (2003–2010), so-called "reality" shows have become soap operas with prizes. The shows have succeeded by rejecting actual scriptwriting for the much cheaper option of casting abrasive people to act the way they think people should act on television. Contestants are put into artificial situations and then encouraged by producers toward conflicts that will cut together well in the editing room.

Shows like *The Real Housewives* (2006–) franchises, the *Basketball Wives* (2010–) franchises, the *Kardashian* (2007–) franchises, and *The Jersey Shore* (2009–) franchises provide the free-base distillation of the soap opera to a set of paradoxically un-self-aware narcissists regularly offending and taking offense, in between product placements. While a wonderful source for examining American culture in the years leading up to the Mayan apocalypse, the shows lack the density of the classic soap operas that were created by actual writers rather than editors.

The scripted female-centered soap opera does make the occasional appearance on prime-time television, most successfully in *The Good Wife* (2009–) on CBS. The female-centered soap is now more likely to be a genre television series featuring a tough woman in the hostile male environment of a cop show like *The Closer* (2005–2011) *Saving Grace* (2007–2010) or blended viewpoint of dramas like *Castle* (2009–). Ironically, the female-oriented soap opera may be best respected in the half-hour cable comedies of female suffering like *The Big C* (2010–), *United States of Tara* (2009–2011), *Enlightened* (2011–), *Nurse Jackie* (2009–), and even *Weeds* (2005–).

Chapter Two

The Singing Detective and British Television

In Which We Consider Dennis Potter's Masterpiece, the Difference Between British and American Television Series, the Novel-Based Series, the Effects of Social Documentary on the Soap Opera and Comedy, and American Variations on Potter's Innovations

The Singing Detective *is the best television I've ever seen.*

—Steven Bochco[1]

The Singing Detective (1986) was a seven-hour television drama unlike anything on American television at the time. It's airing in the United States should be considered at least as significant a cultural shock to American television creators as the arrival of the Beach Boys album *Pet Sounds* was to British rock musicians. Or *Sgt. Pepper's Lonely Hearts Club Band* was to American bands, for those that missed the message going the other direction.

The series tells the story of a writer in a British public hospital ward immobilized by the agony of arthritic psoriasis. The writer, named Philip Marlow, is truly unlikable, shouting misogynistic and racist abuse one minute, whining and sniveling the next. He believes his wife and a lover are conspiring to steal the movie rights to one of his old novels.

Not the stuff of prime-time ratings.

But that's only the starting point. As Marlow is trapped inside his revolting body, the series follows the writer's mental journey through a fevered blending of childhood memories, his detective novel *The Singing Detective,* and his hallucinations, which include the residents and staff of the ward suddenly breaking into lip-synched musical production numbers of popular songs from the writer's childhood. The show is shot in the alternating styles of medical drama, regional history, period detective story, and sitcom, occasionally interrupted by musical numbers.

Perhaps most succinctly, Patrick Malahyde described the series as a psychological case history told as a detective story and set to music.[2]

Definitely not the stuff of prime-time ratings.

The series was screened in cinemas and theaters in the United States, including the Public Theater in New York and the Smithsonian Institution in Washington, DC, received rave reviews in American media and gradually was picked up by individual PBS stations across the country. The series was too controversial for the national PBS operation to purchase the rights for all its member stations.

For many, *The Singing Detective* wasn't television, because the story it told and the way it told it was just so different from anything on American television. It was described as a video novel or as a film on television (not to be confused with a made-for-TV movie).

Steven Bochco, creator of *Hill Street Blues*, was one of many American television creators at the time who were trying to transform network television, and he found in *The Singing Detective* permission to do many things that were just not done in American television.

British Television: Where *The Singing Detective* Came From

One understandable misconception is that *The Singing Detective* came out of nowhere. While it was a revelation to creators and viewers of television in the United States, it was very much a product of British television and the culmination of Dennis Potter's decades of television writing.

British television developed as a government service with a continuing mission laid out in 1922 by the BBC's first general manager, John Reith, that emphasized the service as a cultural tool to elevate the British public, a mission that was also embraced by ITV, Britain's first, regional, commercial network founded in 1955. This included a commitment to staging great theater on television that was expanded by the 1950s to include original works by the cutting-edge playwrights of the present day. These works were expected to challenge the audience with art that educated them while entertaining them.

Theater on television included ITV's *Armchair Theatre* (1956–1974) and *Playhouse* (1967–1983), and The BBC's *The Wednesday Play* (1964–1970) and *Play for Today* (1970–1984), in which television viewers could see classic plays but also work by new playwrights such as Dennis Potter, Trevor Griffiths, Alan Bennett, Alan Plater, and David Mercer. These programs provided venues for new playwrights to develop and showcase their work, but they also habituated the British television audience to the normality of theatrical narrative and experimental themes and techniques on the television.

The result was that even casual viewers were exposed to stylized drama, stories of ideas, and allegories as part of a normal night's television viewing. This seemingly insignificant element had a sweeping effect. If you train an audience in the way American commercial television has, to understand everything they see as a form of realism, they expect to be told direct stories where plot is what matters. (They also, according to Bertolt Brecht, end up trained to accept the lies of entertainment as reality and are easily manipulated by politicians who tell them stories that match the stories they are used to hearing.) If the consensus is that all television should be accessible to children, and should reinforce—rather than criticize—the values held by the audience, the dramas end up black-and-white with simple entertaining characters.

By contrast, if an audience is shown theater on television, they are trained to read a television story on more than a single level. The audience is prepared to consider plots as something other than realism, being habituated to stories that function as metaphor or allegory. They are more prepared for characters that represent competing ideas in a story where neither side obviously wins the argument. Audiences watching theater are also likely to be exposed to heroes who are flawed and villains who are sympathetic.

This does not mean that British television audience did and does not have a large number of people who turn on the telly for a night of mindless entertainment. Just because one is exposed to other approaches doesn't mean that you don't still prefer escapist realism. The letters of complaint that *Monty Python's Flying Circus* (1969–1974) would occasionally use to interrupt their episodes were a regular feature of the BBC's public comment shows and are precisely the sorts of letters that programming such as *The Singing Detective* would inspire. Rupert Murdoch, who was battling the government to start his own television networks, built on viewer criticism of the few seconds of nude buttocks in the first episode of Dennis Potter's show to launch a crusade against the BBC. Murdoch-controlled newspapers became part of a campaign calling for a ban of the rest of *The Singing Detective* for indecency.

Even if it was not appreciated, the exposure to a range of narrative possibilities did provide a vocabulary for British viewers that many American viewers did not have the opportunity to acquire. This might explain the creation of—and the audience for—the absurd comedy of Monty Python or a cult classic like *The Prisoner* (1967) full of surrealist paranoia, political commentary, and psychological puzzles and puzzling.

The loss of original televised theater is often cited as one of the prime causes of the decline of British television that began in the 1980s. Ironically, the other is the expansion of the soap opera form into the other television genres. (Ironic, because the injection of the soap opera into genre television is generally agreed to have revitalized American television at roughly the same time.)

British television was open to more challenging shows for several reasons beyond the BBC's original mission statement. BBC programming was shaped by a different economic model from that of the United States. British television was a government service, paid for by viewer's annual purchase of a license for their television set. The airing of what might be called "quality" programming was possible because the BBC did not have to worry about creating shows that generated large audiences for commercials, which is not to say that the BBC didn't try to provide entertainment to its audience as well. The BBC aired plenty of non-avant-garde programming such as *Are You Being Served?* (1972–1985) and *'Allo 'Allo* (1982–1992), *The Vicar of Dibley* (1994–2007), and *Coupling* (2000–2004). BBC's difference in mission can be seen in the large amount of programming aimed at older viewers—an ignored audience in commercial-driven American television. Just to point to some of the BBC sitcoms of the middle-aged and older that have aired in the United States, *Dad's Army* (1968–1977), *Last of the Summer Wine* (1973–2010), *To the Manor Born* (1979–1981), *Waiting for God* (1990–1994), *One Foot in the Grave* (1990–2000), *Keeping Up Appearances* (1990–1995), and *As Time Goes By* (1992–2005). There are also the edgier comedies *Till Death Us Do Part* (1965–1975) and *Steptoe and Son* (1962–1974, with interruptions), the shows that Norman Lear eventually remade as *All in the Family* (1968–1979) and *Sanford and Son* (1972–1977).

Even with television in place, BBC Radio dominated narrative entertainment into the 1960s, providing drama and comedy long after American radio had turned into a music delivery system. To this day, fiction radio remains vital in the United Kingdom, often written by the same people who write television series, sometimes expanding BBC radio series to BBC television, or vice versa. British television series such as *The Hitchhiker's Guide to the Gallaxy* (1981), *The League of Gentlemen* (1999–2002), *The Mighty Boosh* (2003–), and *Little Britain* (2003–2006) all began on radio. Most of the creators of *Monty Python's Flying Circus* (1969–1974) worked back and forth between radio and television. More recently, *Torchwood* (2006–) fans heard those stories continue on BBC Radio between television seasons. Both shows and writers move between the two media without comment.

The Importance of the Everyday

One other unlikely though significant influence on British television drama was the BBC's commitment to documentary from the earliest days, growing out of government support for what were called "positive propaganda" films going into World War II. The idea was developed by Edward Grierson, the man who came up with the name "documentary" (referring to Robert Flaherty's *Nanook of the North of 1922*), who later helped shape the creation of Canada's Film Board and television.

Grierson had promoted the idea that, in response to Nazi propaganda, England should produce films to show the English how well their society functioned, that is "positive propaganda." The most famous of these Grierson Unit films is probably *Night Mail* (1936), a dramatic documentary about how the mail gets from London to Edinburgh on an overnight train. These films, which ran in movie theaters as part of an evening's entertainment in the 1940s and 1950s, set the model for the social documentary movement, a type of documentary generally viewed as the British contribution to the field of documentary, about everyday people and everyday life that became a staple of BBC programming that has no parallel in American television, although Canadian television followed the same model.

Such documentaries have influence beyond their subject matter. They habituate audiences to seeing "normal" people on television and seeing stories about the uniqueness of a "normal" town or neighborhood, a group of people, a town's marketplace, or a business, or a day in the average life of a bus conductor, as normal television subject matter. By setting a premium on the drama of the everyday, British television taught viewers to see the drama in the uneventful story (by Hollywood standards) as opposed to the celebrity-based big city story with great dramatic characters accomplishing great things. That principle shapes a large part of the dramas and comedies that end up on the BBC. The audience learns to read the drama and comedy of the small and the slow in a way U.S. television doesn't encourage.

At the most basic level, this orientation in British television tells the audience that normal is a shopkeeper from Shropshire or an immigrant custodian in Luton, whereas American television tells its viewers that normal is Jennifer Anniston playing an unemployed college graduate able to live in Manhattan without economic concerns, that George Clooney is an average tormented doctor in Chicago, and that normal America is celebrities depicting characters from New York or Los Angeles.

One place the effect of this difference can be seen is in the dominant form of the soap operas that developed on British and American television. American soap operas followed the model of the "women's weepies" from Hollywood film, telling stories of heightened melodramatic conflict. By the 1970s, the stories of American soaps had a broadness of character and situation with near-caricatured beautiful, well-off, and high-strung people and absurd plot twists to keep the audience in a heightened state of anxiety. By contrast, British soap operas followed social documentary in tending toward a more low-key naturalism with a focus on everyday characters, rather than catalog model wealthy, the joke about British soaps being that nothing ever happens on them. Over time, the American model came to influence the British soaps, but changes are relatively recent.

Most recognizably for American readers, the BBC social documentary tradition gave us the situation comedy, *The Office* (2001–2003). Ricky Gervais and Stephen Merchant's original British series about a small paper company from Slough pretends to be one of the dull BBC social documentaries. The underlying joke of the show is that Gervais's character, David Brent, doesn't see himself as a normal person in a dull documentary, he sees the presence of a documentary camera as his opportunity to become a television celebrity.

That British television joke doesn't translate to American viewers because it is simply the truth of American "reality" shows, nonentities trying to turn themselves into celebrities through the presence of a television camera. The American version of *The Office* (2005–) kept the documentary, handheld form of the show but quickly dropped the redundant underlying joke.

The Short Series

The BBC model for a television series is very different from American television. One of the reasons there could be much more experimentation is that a British series is generally six or eight episodes. If the show is successful, the creators might be asked to make another. American cable series have increasingly followed the British model.

A series in the British model was generally created by only one or a pair of writers, with the assumption that the series was a "one-off," so the show could be written in a theatrical or novelistic mode, developing characters and story across the episodes, safe in the assumption that audiences would see the shows and follow the story toward a planned ending in the final—sixth or eighth—episode. In such writing, there was more of an assumption that the story would follow classic theatrical and novelistic dramatic structures, with a character or characters transformed by events over time.

By contrast, American network television series are created with the assumption that they will run forever. This leads to many differences in writing. American series are written with characters remaining unchanged and each episode capable of standing on its own, so that first-time viewers will enjoy any episode they watch and so casual viewers can watch a few episodes out of order and still enjoy themselves. When we discuss American television authorship, we generally mean that the creator and executive producer of a show write the first episode and rough out the story, which is then actually written by individual staff writers and then rewritten/polished by a room of staff writers under the guidance of a story editor.

Authorship in British television is generally an individual or team experience compared collective experience in U.S. television. Confusion might arise from the fact that Americans tend to refer to a series as being the work of a single creator—Steven Bochco, Michael Mann—even when they are often co-created and group-written. There are exceptions but, generally, the creator of an American series writes very few of the actual episodes. There's nothing deceptive or inferior about this practice, in fact some argue that collaborative writing is superior to solo work. Hollywood film writing in the studio era was a collective process and that model transferred naturally to network television.

Novels on Television

Another successful form of television narrative for the BBC was the adaptation of, generally historical, novels for television. The adapted novels again reinforced the ideas of television as a closed narrative form, suggesting that a story—in this case a television series—had a beginning, a middle, and an ending, during which viewers watched a dramatic arc played out across a period of time as characters were confronted with a conflict that led them to a dramatic transformation by the end of the story.

These British adaptations of novels, sometimes known as "heritage" dramas, ended up airing on public television stations in the United States, generally within the framework of the *Masterpiece Theatre* franchise. John Galsworthy's *The Forsythe Saga* (26 episodes, 1967), Robert Graves' *I, Claudius* (12 episodes, 1976), John LeCarre's *Tinker Tailor Soldier Spy* (7 episodes, 1979), Evelyn Waugh's *Brideshead Revisited* (11 episodes, 1981), LeCarre's *Smiley's People* (6 episodes, 1982), and Mervyn Peake's *Gormenghast* (4 episodes, 2000) offer just a sampling of the most successful adaptations of novels that trained British audiences to the idea that, even on television, stories had the properties of novels.

The successful adaptations of novels set the model for adapting mystery novels and later short stories with the Sherlock Holmes and Poirot, Inspector Morse, and Rumpole of the Bailey series that also ended up rebroadcast on PBS's *Mystery* franchise or running independently on local PBS stations (along with "novelistic" historical series packaged exactly like the novel-based series, such as *Upstairs, Downstairs* [1971–1975, 2010] and *Downton Abbey* [2010–]).

The rebroadcast of the literary adaptations from Britain on public television in America led many American viewers to see novel-based television as boring and educational. The form did not cross over to American network television, with one exception.

In the United States, there was a roughly decade-long burst of miniseries from the mid-1970s that attempted to adapt the model of the novel adaptation to American television. This meant a more expensive production, often discussed as being more filmlike in its execution, and an airing on succeeding nights rather than once a week. The most famous and influential was the adaptation of Arthur Haley's *Roots* (1977), but there was also James Clavell's *Shogun* (1980) and Leon Uris's *QBVII* (1974). Later nonadapted miniseries dealt with "issues" and took a novelistic form, most famously *Holocaust* (1978) (which included a performance by a young Meryl Streep) and *The Day After* (1983), about the aftermath of a nuclear explosion in the United States.

British television's emphasis on the short series, the tradition of series written by a single author, and the success of novels adapted into television series—like the seven series of *All Creatures Great and Small* (1978–1990) based on James Herriot's first two novels *If Only They Could Talk* and *It Shouldn't Happen to a Vet*—all encouraged the writing of series by novelists. Both dramas and comedies based on novels and written as series by a single novelist continue to be a staple in British television: comedies such as *The Biederbecke Trilogy* (1985–1988) and *Oliver's Travels* (1995) by Alan Plater, the trilogy of the *Reginald Perrin* series (1976–1979) and novels by David

Nobbs, and, most recognizable to Americans, the police procedural series *Prime Suspect* originally written by novelist Lynda LaPlante.

The Americanization of Jane Tennison

Prime Suspect (seven series, 1991–2006) was a stunning popular success when rebroadcast on public television in America. Each series was written as a novel for television about a police inspection focused on a prime suspect, with Inspector Jane Tennison struggling against all manner of fallout from institutional and societal sexism against a woman in authority. Two of the *Prime Suspect* series were written by series creator, novelist Lynda LaPlante. The others followed the single-author model.

Each series (or season, in the United States) would follow the same model of a single case, in each case the personal struggle of the admirable but not always likable Tennison is exacerbated by the levels of controversy associated with the case.

The Closer (2005–2012) was a successful attempt to adapt the themes of *Prime Suspect* for American television. Comparing the two shows reveals the continuing difference between U.S. and UK television even at the beginning of the 21st century.

Primarily, there is no continuing season-long case on *The Closer*, each episode closes with a case resolved and a personal closure for Deputy Chief Brenda Leigh Johnson (Kyra Sedgwick). In early seasons, Chief Johnson can get snippy but manages witnesses, suspects, her men, and the system through her application of Southern woman's charms. Chief Johnson is likable to almost everyone on the show and to the viewer. There is an air of drama with light humor in most of the episodes, often based on the gendered conflicts, and Chief Johnson's inherent superiority to the people who underestimate her because she is a Southern woman.

This is a stark contrast to Helen Mirren's Inspector Tennison, who ends up winning the audience over to her side but is often resentful, angry, petty, develops a debilitating drinking problem, and is rarely likable, even when she tries to be nice. She wins the respect of some of the men she commands but clearly rejects the sorts of flirtatious charm that smooth the rough edges for Chief Johnson in *The Closer*. Most different, *Prime Suspect*'s tone is bleak and there is always a real threat hanging over Tennison's head, as opposed to the American use of light comedy.

Both shows accomplish what they set out to do and provide their audiences with the experience they expect. Even as the final season of *The*

Closer turns darker, as Chief Johnson is unfairly sued by a malicious lawyer, the American show is a reinforcement of episodic entertainment and the triumph of the good guys over injustice. The British is a bleak novelistic treatment of a murder investigation with no comic relief.

In the fall of 2011, *Prime Suspect* was officially adapted as an NBC series. This translation of the British series was promoted as an episode-based series about an opinionated cop we can like, a female variation on *House*, a witty abrasive female cop who gets things done. Where Mirren's character was abrasive, Bello's character is "sassy." Where Mirren was a boss put in charge of a group of male detectives who resented her, the American version has Bello as the only woman in an all-male detective squad, with her male boss defending her. Promotion for the new series stressed that there will be a mystery solved in each episode, referring to the stand-alone episode structure as "an American-style procedural."[3]

The great distinction from Helen Mirren's detective is that Bello's character gets to win—solving cases but also winning office battles, cases, struggles with her father, her sister, her boyfriend's ex-wife, witnesses, suspects, taxi drivers, and strangers multiple times in an episode. Even when she nominally loses, she still turns the situation so that she comes out on top.

The British series is one long case full of awkward setbacks and struggles so that when the series ends there is a real question about whether the final result was worth what it cost Tennison and others changed by the case. The American series regularly gives the fantasy of control, sexiness, attitude, and success tied up with a feel-good ending. One challenges the viewer, the other tells a bedtime story the viewer has heard before. Both series do what they intend to do and they do it very well, although the American *Prime Suspect* failed to attract an audience (possibly because it aired on the cluttered Thursday night). The difference is in what we expect a television series to do.

The Singing Detective

The Singing Detective (1986) tells the story of a paranoid and hallucinating writer incapacitated in a public ward with horrific arthritic psoriasis. As Marlow, the writer, suffers he struggles to solve a mystery of his marriage, his novel, his childhood memories, and his hallucinations as each level of reality bleeds into the other.

At the heart of the series is a portrait of a complicated, off-putting, when not completely unlikable, character. Michael Gambon's performance as Marlow, the detective novelist, is integral to the script working, as the character begins as physically repulsive, raging, and weeping and hallucinating by turns, about as far from a television leading man as might have been imagined at the time. The fact that it is based in a self-portrait by the author is part of the caustic force of the series; it sits somewhere between a confession and a psychic disembowelment.

Potter made two other series related through their use of song and consideration of the disconnection between types of consciousness, stories, and memories—*Pennies from Heaven* (1978) and *Cream in My Coffee* (1980)—but *The Singing Detective* is clearly the culmination of his life's work.

Until the 1970s, the BBC was very much a writer's environment, where series were generally shot on sets in-house, and writers could have a hand in every stage of productions. But as the 1980s arrived, there was a move to filming on location and with it a cinematic model that saw the power of the writer eclipsed by the control of the director. An even greater impact would be felt from the shift in government attitudes toward the BBC as the Thatcher government came to power. As Kenith Trodd, the producer of *The Singing Detective,* explains on the commentary track of the DVD set, the Potter series was made at the end of the BBC's creative period. There is no way that the series could be made at the BBC today, or anywhere, except possibly HBO.[4]

Even for British television, *The Singing Detective* did not behave like television. The series' use of music was probably most influential, but it was the attempt to tell a story subjectively, with all that entails, and its use of an unattractive "hero" that had the greatest effects on future television. The other influential aspect was its varied uses of self-referentiality, from a scathing self-examination of a fictional version of himself to Potter's deconstruction of popular culture, taking apart the form and meaning of sitcoms and television drama, as well as detective fiction and popular song.

The show is a multilayered excavation of the writer, both in the juggling of storylines and realities, but also in the story being about the imagination of an author under assault by those who care about money, not the art, not the artist. As the story reveals that Marlow's detective novels grow out of the personal crises of his childhood, many of those details also arise from Potter's own childhood in the West Country Forest of Glenn, and his debilitating psoriasis, and the religious themes of his earlier work.

Potter demonstrated television drama as not just an auteur art but as a fictionalized autobiographical art. *The Singing Detective* showed that a television program did not have to be part of a factory assembly line of narrative but could be personal. It could be self-referential—drawing attention to the conventions of television genres—and it could be stylized far beyond the American network ideal versions of realism.

Within the novel-within-the-novel, the Singing Detective taunts a client with the statement that “there are songs to sing. There are feelings to feel. There are thoughts to think. That makes three things. And you can't do *three* things at the same time.”[5] But Potter is actually taunting the viewer with a description of exactly what the program is doing. *The Singing Detective* forces the viewer to shift gears between intellectual, emotional, and aesthetic pleasures, eventually breaking down the boundaries so that the different reactions leak into each other. American television offers those experiences, but generally well-separated and in moderation. The complicating of our reactions is central to Potter's strategy.

It also told a story that was confusing, with events that turned out not to be what we thought they were at first exposure, with a fairly repellant lead character who alternates between self-pity, paranoia, and rage, drifting between reality, his novels (in which he becomes his singing detective), his childhood, his fantasies, and bizarre hallucinations in which people in the real world suddenly begin lip-synching to songs from his childhood.

These moments of dislocation are both comic and horrific as we—and the character of Marlow—try to make sense of them. As with classic musicals, the songs express the unspoken feelings of the lead character, but Potter's use of songs fractures reality in a nontraditional way. The songs are familiar, safe, a bit twee—even to American audiences—but their appearance is threatening.

What Potter does goes far beyond the previous limits of music in television narrative that were established in *Rock Follies* (1976–1977), the story of the formation of an early all-female rock group; Roxy Music's Andy Mackay wrote wonderful original songs for the band that commented on their situation implicitly when performed during gigs within the continuing story but then became an explicit commentary on the themes of the episode when re-performed outside the story at the end of each episode.

Potter's use of music is only part of the show's strategy of de-familiarizing the recognizable to wrench apart any clean response viewers can have. In 1986 when this appeared, irony was not yet the default cultural position. Combining the unresolvable juxtapositions with a story whose reality kept changing—long before *The Matrix* (1999) or *Inception* (2010) mainstreamed the idea of malleable reality—created narratives where the audience could not rely on its perceptions. Viewers also couldn't trust the conventions of a television story, because if they appeared at all, it was only to be violated.

Like the music, the different sections of *The Singing Detective* take the form of recognizable popular entertainment genres but don't provide what the audience expects. The supporting characters of the ward scenes come from British sitcom conventions of the day, the musical numbers are surreal interpretations of musical production numbers, and the sections of the detective novel-within-the-novel are shot in the style of a BBC period mystery, with *the Singing Detective* dropped in.

Beneath this use of genres was the idea that internal reality creates our understanding of external reality, that what we perceive is simply the result of what we feel, not in a Freudian sense, but in a broader sense of humans making sense of the world through the media that they consume. Marlow makes sense of his life through the clichés of the hard-boiled detective novel, with a bit of the postwar espionage drama thrown in, and then the supposed escape into the nostalgia of childhood popular culture.

In Marlow's childhood, popular entertainment took the form of local men singing hit songs on the pub stage. Having witnessed his mother in an adulterous act, he turns to God and gets no answers so, as a child, he decides he will become a detective. As an adult, he blends the two into his fictional alter ego, the cool and commanding Singing Detective who does not get hurt and is always in control. But that fictional version is not enough, the childhood pop songs now torment him rather than comfort him and the hard-boiled narratives shape the way he sees his wife, as a femme fatale, as he lies helpless in hospitalized agony.

The popular culture is inseparable from the personal, especially for the writer. Named Philip E. Marlow, the relocated *E* being the only thing to separate him from Philip Marlowe, the American hard-boiled detective created by Raymond Chandler, an English boy transplanted to America who made good in Hollywood by writing novels and films of manly men who solve mysteries.

One of the ironies of the detective novel that lies at the heart of *The Singing Detective* is that it is generic tripe, a postwar British knock-off of the actual Philip Marlowe stories (though the misogyny seems more Mickey Spillane). Chandler's original Marlowe (like Chandler himself) is actually an isolated alcoholic tormented by betrayal and failure; even when he succeeds, he has no lasting effect. The only triumph of Chandler's Marlowe is to be the one guy who understands what happened. By contrast, Marlow's Singing Detective is the little boy's ideal of an all-seeing, all-powerful man—that is, Potter's surrogate in the show is a writer of infantile escapist entertainment, not great art.

Potter's series goes beyond simple self-referentiality into a depiction of the way that identity is a loose mesh of competing personal and commercial fictions, each viewed through the lens of the other. The struggle is not, as Freud would have it, to be able to tell a coherent version of one's life but to get those pieces to cohere into a functional blend of reality and fiction. Marlow does this at the end of the series, essentially healed by embracing the reality of the pulp escapism.

Dennis Potter was once described as having not a "love-hate" relationship but a "hate-hate" relationship with Hollywood. He desperately wanted to be part of it, even as he savagely attacked the Dream Factory fantasies in his work. One of the strangest historical footnotes to Dennis Potter's television triumphs is that he wrote and sold his own film adaptations for both *The Singing Detective* and *Pennies from Heaven*. While both movies have great moments and carry something of the message of the television originals, they are softened and conventionalized in a way that mainly demonstrates how right Potter was about the unique power of television drama.

The Singing Detective's Influences on American Television

It's easy to overstate the direct influence of Potter's one work, but American television shows continue to pass through the doors that *The Singing Detective* opened in a way that echoes the influence of an album like *Pet Sounds* on decades of later concept albums. Or Alan Moore's graphic novel *Watchmen* that changed people's ideas about what comics can say but also how they can say it. One dense work can influence in various ways, mainly by giving permission for later artists to go further into nontraditional territory.

We can see the consequences of the permissions people took from *The Singing Detective* echoing down through 30 years of American television, including plenty of work by people who don't know what *The Singing Detective* is but do know *St. Elsewhere* (1982–1988) or *The Sopranos* (1999–2007) or *Scrubs* (2001–2010).

For some, like Steven Bochco, the influence was specific in the use of music on television but also general, in the permission to explore longer story arcs, more psychological and emotional realism, more realistic treatment of sex, more self-referentiality, and more meddling with the institutional forms of television genres.

For some, it was the idea of a television show built around an unlikable, if not offensive, hero with a complex character. A character the audience could neither like nor dismiss. Such characters are common in literature and theater but were essentially nonexistent on U.S. television. After *The Singing Detective*, we find disturbed heroes with attractive villains stealing the show on *Wiseguy* (1987–1990) and *Murder One* (1995–1997), then unlikable heroes on *Profit* (1996), *The Sopranos, Deadwood* (2004–2006), *House* (2004–2012), and *Breaking Bad* (2008–2012).

The Singing Detective demonstrated to Americans that a television series can be a deeply personal artwork, reflecting the vision of a single creator. Network television had a reputation for collective mediocrity. Potter offered the model of the writer-creator as television auteur.

These days we tend to connect television shows to their individual creators, even if they don't, as David Milch is reputed to, dictate every line of the scripts to their writing support team or write the scripts out in longhand, as David E. Kelley does. David Simon of *The Wire* (2002–2008) and *Treme* (2010–) is the most recognizable television auteur, even though he regularly credits and praises his team of writers. It is David Chase's *The Sopranos* and Joss Whedon's *Dollhouse* (2009–2010), *Firefly* (2002–2003), and *Buffy the Vampire Slayer* (1997–2003). While these identities were known within the television industry, they are now known to the public and most longform television is attributed to a single authorial voice.

For others, the permission to use the autobiographical became key points. This altered autobiography is a basis of modern literary fiction, especially the clichéd first novel. It also inspires some plays and fewer feature films, perhaps because each medium moves further from the control of one creative voice. Potter's taking of a novelistic sensibility into the teleplay offered permission to use genre television forms in the service of a more personal exploration.

Bochco's *Doogie Howser* (1989–1993) was a tribute to his once child prodigy father. *NYPD Blue*'s (1993–2005) Andy Sipowicz is a self-examination through a fictional portrait of David Milch's father. David Chase's terrible relationship with his mother (or his relationship with his terrible mother, depending on your grammatical preference) became the heart of Tony Soprano's relationship with his mother, Livia, who tried to kill him before the end of the first season. Roseanne Barr constructed a version of what her life could have been. *The Larry Sanders Show* (1992–1998) becomes the alternate reality Garry Shandling would have lived if he had taken the offers to host a talk show when Johnny Carson retired. At a minimum, *The*

Simpsons (1989–) family members share their names with the family of creator Matt Groening. Larry David and Jerry Seinfeld create a fictional version of themselves on *Seinfeld* (1990–1998) and then Larry David goes on to create a sort of "reality sitcom" about himself in *Curb Your Enthusiasm* (2000–).

The Singing Detective's radical idea of depicting reality, dream, hallucination, and memory as overlapping and cross-influencing states within a television narrative has become a minor convention of television shows. American television has done a wide range of variations and simplifications across genres. By the second season of *The Big C* (2010–), when Cathy dreams she is onstage with a magician played by her cancer specialist, it's practically a cliché. *The Sopranos* famously used dreams, mainly Tony's, to show us realities that the character had not yet admitted to himself. Since *The Singing Detective*, the convention of a show opening in apparent reality that turns out to be a dream/hallucination regularly appears in shows like *Rescue Me* (2004–2011), whose final episode begins with a funeral service that turns out to be a fantasy, flashbacks occur without warning, followed by drunken violence that turns out to be a premonition of the consequences of a drink that is never taken, only to end with a conversation with a ghost.

Lost (2004–2010), *Fringe* (2008–), *Wilfred* (2011), *NYPD Blue, The Days and Nights of Molly Brown* (1987–1991), *Buffy the Vampire Slayer, Community* (2009–), *Scrubs, House,* and *Battlestar: Galactica* (2004–2009) are just the first examples that come to mind of the use of the device of reality revealed to be unreal. *Wilfred, Scrubs,* and *Community* use the alternate reality device and shifts between genres for their pure entertainment effect. *Wilfred* uses the plot device self-mockingly as the alternate reality is represented by a talking "dog" that is an Australian man in a cheap dog suit and a painted nose.

Proof of how conventional the supposed shock effect of these alternate realities has become can be seen in the way that *Breaking Bad* has used audience expectation to mislead them. Several episodes of *Breaking Bad* begin with the stark visuals and absence of dialogue that are now clichéd cues that the audience is seeing a dream, only to have the scene turn out to be a very real, and disturbing, reality.

The alternate reality convention is the central device in the three seasons of *The United States of Tara* (2009–2011) where Tara regularly becomes one of her multiple personalities and "normal" Tara must confront and negotiate with her alters as if they were just other characters in the story, all on the way toward solving the psychological mystery of what caused her to fragment herself to survive as a child.

Roseanne (1988–1997) was a sitcom on which Roseanne (the show's creator and star) continually attempted to push the limits of what was acceptable social realism for prime-time television. In the final season, the character Roseanne won more than a $100 million in the lottery, and the family spent the season living a life "like the people on television." But in the final episode, Roseanne explained to viewers that her husband, Dan, had died a year earlier and the whole ninth season had been her escape from her depression into a popular culture fantasy. She explained that she had started writing about her life—writing that is, apparently, the entire television series—as a way that Roseanne (the character) tried to make her life better in fiction than it was in reality. The result is an after-the-fact reworking of the meaning of the entire series in the last episode, turning *The Singing Detective*'s tormented confusion of fiction and reality in trauma into an act of self-healing from trauma by rewriting one's life as genre fiction.

Even with *Rescue Me*'s extensive use of alternate realities, *House* has probably wrung more variations on *The Singing Detective* than any other American show. *House*'s Emmy-winning episode "Three Stories" is purely *The Singing Detective* without the music, as the crippled doctor tells three case histories to a class of diagnostic students. As viewers see the stories, Dr. House is describing the identities of the characters and the actors playing them (including Carmen Elektra playing herself) repeatedly shift between the three stories and within the individual stories until the viewer gradually recognizes that one of the three stories is about House himself, and it becomes the emotional and pivotal story of how the doctor lost the use of his leg and the woman he loved.

In the two-part fourth season finale, House spends two television hours trying to remember the identity of the person he knows needs help after a bus crash. House talks to the characters in his subconscious, restaging, and repeating scenes, as he gradually comes to realize that he is responsible for the death of his best friend's fiancé.

Most sweepingly, at the end of fifth season, viewers learn that at least some of what we have seen throughout the season has been Vicodin-created hallucinations that House has incorporated into his reality. Viewers have received hints that something is off, but not until the final episode is it clear that significant events they watched never "really" happened. In both cases, we see House's different levels of reality bleeding into each other in a natural development from Potter's more theatrical depiction.

Music on Television

Music in television drama—often in hallucinatory contexts—is one place where Potter's influence is pervasive but hasn't aligned with his intentions. Potter overtly rejected the logic of the musical, dissecting the form through scenes of lip-synching of popular tunes of bygone eras, exploring the important meaning—often two-sided—that popular songs have in our lives.

American television's use of music in drama tends to have avoided this critical function to simply provide music. There are a few exceptions that use music in the same way as Potter did; the use of "Nights in White Satin" at the end of the first long-arc story of *Wiseguy*, *House*'s "Forgot Your Troubles, C'mon Get Happy" hallucinatory song-and-dance number in an operating theater, and possibly the shows repeated use of "You Can't Always Get What You Want," but they are few. *Battlestar: Galactica*'s use of Bob Dylan's "All Along the Watchtower" as the brain-implanted Xylon wake-up call that brings about the final closure of the series and becomes a sort of cosmic lip-synch, only to become the final words of the series played over an eon-sweeping final montage is probably the use of the song comes closest to Potter's. Just as in *The Singing Detective*, we see the recontextualization of a well-known popular song by using it in a way that invokes its original meaning, gives it a specific new meaning within the show, and sets up an dissonance between the different meanings that the audience must resolve on its own.

More common Potter-inspired but not Potteresque uses of music include the failed series *Eli Stone* (2008–2009) in which a lawyer having a midlife crisis begins to hallucinate people breaking into pop songs, with George Michael appearing in several episodes as both an hallucination and later as himself. The singing was a sign of Eli's brain tumor or Eli's prophet vision.

Generally, the significant uses of music in American television drama have been of two kinds. The music can be recognizable songs used to enhance or comment on a scene—best exemplified by the use of "Something in the Air Tonight" in the pilot of *Miami Vice* (1984–1990), which spearheaded the use of hit songs on television soundtracks. *Northern Exposure* (1990–1995) probably did this most artfully, with music played diagetically on the local radio station providing a refreshingly ambiguous commentary on the themes of the episodes, in the same way that original poetry was used on *Oz* (1997–2003). *WKRP in Cincinnati* (1978–1982) had the DJs play actual hit songs when they were on the air, to reinforce the reality of the show's rock radio station. For economic reasons, this practice settled into the turn of the century convention of using a single popular indie hit per episode, generally over a meaningful montage, as seen on shows from *Miami Vice* to *Grey's Anatomy* (2005–).

The other convention is the musical, at which *Cop Rock* (1990) famously failed. Since then, there has developed something of a convention of the single, "very special" musical episode in an otherwise tone-deaf show. As with *Cop Rock*, the shows tend to use original songs—not standards—although they do seem to follow the *Singing Detective* rationale of altered consciousness leading to musicals.

The cast of *Buffy the Vampire Slayer* became magically, musically possessed for one episode. The cast of *Scrubs* were heard singing by a patient suffering from a consciousness-altering syndrome. The most unnecessary use of the device occurred on the "Britney Spears" episode of *Glee* (2009), where some of the characters had musical hallucinations under the influence of dental anesthetic, performances that were only slightly different from the numbers they performed during the nominal reality of the show.

Otherwise, *Glee* seems to have proven that American dramatic television is most comfortable with its songs familiar and grounded in realism. While the show slides into surreality or imagination, it is apparently forgiven by the viewing audience since everyone involved is a singer in the reality of the show. In this, *Rock Follies* may be the British musical series that comes closest to American tastes, a judgment confirmed by the success of the similarly "realist with moments of surreal performance" 2012 NBC show *Smash*.

The closest application of Potter's approach to popular music is actually found in the use of film and television clips in the HBO situation comedy, *Dream On* (1990–1996). The conceit of the show, presented in the opening titles, is that Martin Tupper (Brian Benben) was raised by his television set with the result that any time he reacts to something we see his mental process through quick clips from classic black-and-white film and television shows.

Generally the clips are out-of-context one-liners, closer to the almost subliminal clips of Reginald Perrin imagining a hippopotamus every time his mother-in-law is mentioned on the British series *The Fall and Rise of Reginald Perrin*. Occasionally, the mediated flashback bleeds over present reality, as in the sixth season episode "9 1/2 Days," where Alfalfa from the

Little Rascals is seen singing "Love in Bloom," a serenade that continues over a montage of Martin and an etiquette expert having vigorous and varied sexual relations. In another episode, "A Midsummer Night's Dream On," Martin's ex-wife complains of having had only two penises in her life. We cut briefly to Bob Hope saying "two heads are always better than one," and then Martin comforts her.

Where Potter's songs of the past do call back his childhood innocence, they also take on added layers of ironic or foreboding commentary and are a reminder of his hallucinatory state. By contrast, *Dream On*'s fragmentary flashbacks may suggest Martin's emotional state, what he'd like to say or they just make a sarcastic editorial comment that seems to come from the show more than Martin's subconscious, without any deeper invocation of childhood innocence.

But then, *Dream On* is simply going for simple laughs, not trying to make any broader dramatic statement or trying to suggest the complexity toward which *The Singing Detective* was aiming. This postmodern use of juxtaposition without deeper context works very effectively as a comic device and ends up further transmuted into the generic fantasies of J.D. on *Scrubs*, which grow out of popular mediated realities but don't actually quote a specific source.

In an interesting echo of *The Singing Detective, Dream On* begins with Martin being divorced by the wife he still loves and the series ends with him remarrying her.

The generic self-awareness of the Oliver Stone/Quentin Tarrantino film *Natural Born Killers* (1994) gives us the story segment as sitcom. Rodney Dangerfield plays Juliette Lewis's sexually abusive father in a flashback depicted as a laugh-tracked situation comedy. Generally situation comedies like *Community* and *Scrubs* and *30 Rock* (2006–) have used this element of juxtaposed television genres most effectively.

Most pervasively, the juxtaposition of genres is used in a less disruptive way than Potter did in *The Singing Detective*, generally used to create successful longform television, with one genre providing the long arc and the other providing the episodic payoff.

This transformation began before *The Singing Detective*, and can be seen most prominently in *Hill Street Blues* and the other series that blended an existing television genre with elements of the soap opera to create the long arc. But perhaps the most influential idea in *The Singing Detective* is that you can blend genres to create a new statement. This, we will see later, is central to the development and sustainability of longform television of the DVD novel. It is seen in longform shows as diverse as *United States of Tara*, *Gilmore Girls* (2000–2007), and *The Sopranos*, *House*, and *Veronica Mars* (2004–2007).

Most important to this discussion, *The Singing Detective* showed that television could be made that transcended television and would live on well past its airdate to be considered a form of literature.

Chapter Three

Steven Bochco and Too Many Stories, Not Enough Time

In Which Steven Bochco's Attempts to Change Television are Considered, along with the Soaping of American Television, Some Early Rumblings of Dramedy, the Single-Story Season, and American Musical Television

"We are dying," says Mr. Bochco flatly of dramatic television on the networks. He predicts the future for TV lies increasingly in "some form of pay television," such as cable and video, and suggests, "I think that will be liberating."[1]

—*Baltimore Sun*, February 12, 1992

Steven Bochco famously said that *The Singing Detective* (1986) is the greatest television show ever. Generally, that is interpreted to mean that, since Dennis Potter's series had singing, *The Singing Detective* is the reason that Bochco made his widely criticized musical cop show, *Cop Rock* (1990).

That's unfair to both of them, because it overlooks Bochco's interest in Potter's very personal way of creating television shows that were the equal of theater or literature in the themes they considered and the permission he found to bend the rules of American television that Potter broke in Britain.

Bochco's praise of Potter came at a joint appearance they made in January 1992, honoring Dennis Potter at the Museum of Television and Radio in New York. At the time, Bochco had already fundamentally changed American television with his hit series *Hill Street Blues* (1981–1987) and had two hit series running, *L.A. Law* (1986–1994) and *Doogie Howser, M.D.* (1989–1993). He was a very successful creator of American television and understood the freedoms and the constraints of the form.

Bochco saw in Potter's work models for making television as art, but it is important to remember that Bochco also saw American television as a medium for entertainment, not especially amenable to art.[2] This distinction often gets lost. Bochco's career shows no evidence he was trying to *copy* Potter, but he did take permission from Potter's successes to expand the dramatic vocabulary within American network television.

Potter's example encouraged Bochco to break down the boundaries between television genres, to mix absurd comedy with drama, to move toward stylized and subjective television (including trying musical drama on television), and to question the conventions of episodic television (including using network television as a medium for more autobiographical exploration). Potter's greatest works also made a compelling case for longform television drama, instead of creating a series of stand-alone episodes. Bochco took inspiration from Potter's work to do more of what he was already trying to do.

Steven Bochco was born into an artistic New York family, his mother a painter and his father a concert violinist. His childhood pointed toward a career in the arts and, after studying to be a playwright at Carnegie Tech, he moved to Universal Studios in the mid-1960s where he spent 12 years writing episodic dramas like *Ironsides* (1967–1975) and *McMillan and Wife* (1971–1977), learning the business as his playwright training was reshaped to create compelling formulaic narrative for television. Bochco's writing for *Columbo* (1971–2003) may be his most recognized achievement of the Universal years.

Hill Street Blues: More Than a Cop Show

Hill Street Blues (1981–1987) is widely credited with reviving the cop show genre on American television. Its success has been variously attributed to Robert Butler, the director of the pilot episode who set the style of the series, to the talented group of writers—including Anthony Yerkovitch, later co-creator of *Miami Vice* (1984–1990), Mark Frost, later co-creator of *Twin Peaks* (1990–1991), Dick Wolf, later creator of the *Law & Order* franchise (1990–), and David Milch, who we will get back to in a later chapter—and to the network for allowing the show to continue despite low ratings (and then for recognizing that the lower overall ratings were less important than high ratings in key demographics, especially those in large midwestern urban centers, as network television adjusted to the new realities of competition from cable).

But most often, its success is attributed to Steven Bochco, and the series is discussed as a Steven Bochco show, his first big hit.

Hill Street Blues changed the way American television dramas were told, without the radical assault some found in Dennis Potter's work.

In his series, Bochco never left the basic vocabulary of network television. His achievement was in finding ways to reassemble elements of genre television to create something new; longform American television with multi-episode story arcs and season-long stories with individual episodes that could still satisfy the first-time or sporadic viewer. *Hill Street Blues'* narrative structure quickly became the norm in American television.

Bochco revived the cop drama by blending it with the soap opera. *Hill Street Blues* offered the formula of the cop show genre but expanded the points of emotional connection by illuminating the off-duty lives of the police officers, drawing them out as not just "cops" but as feeling human beings.

According to an interview with Bochco that appears in Pamela Douglas's *Writing the TV Drama Series*, longform television was a side effect of the number of stories he wanted to tell.

> When you end up creating a show with seven, eight, nine character—in response to that, ask yourself how can you appropriately dramatize that many characters within the framework of an hour television show? And the answer is that you can't. So you say, okay, what we have to do is spill over the sides of our form and start telling multi-plot, more serial kinds of stories. . . . The show began to dictate what it needed to be. Probably the smartest thing that Michael [Kozoll] and I did was to let it take us there instead of trying to hack away to get back into the box. We just let it spill over.[3]

The large ensemble cast with multiple storylines was not new to American television but it was seen mainly in the daytime dramas Americans call soap operas. One of the contemporary innovations in nighttime programming had been the introduction of the nighttime soap operas such as *Dallas* (1978–1991) and *Knotts Landing* (1979–1993). These shows took the daily melodramas of daytime television and made them weekly extravaganzas of conspicuous consumption. These continuing stories of fabulously wealthy people betraying each other while still unable to find happiness became successes of excess.

Some attributed their success to an audience among working women who couldn't watch daytime but wanted to continue to consume their serial melodramas. Others saw the shows more as a desire for what Tom Wolfe called "plutography," the pornography of obscene wealth. Others stressed the appeal of flamboyantly overheated melodrama.

Whether intentionally or not, Bochco picked up on some of the less exaggerated elements of the daytime dramas—the suffering of characters who care too much, the suspicions and betrayals of relationships in a world where all relationships are "relationships"—and used them to turn a cop show into longform drama.

Hill Street Blues made one significant shift from traditional soaps; the traditional focus on the suffering of women in settings of nurture was replaced by a shift to the male sphere, exploring the suffering of men in the non-nurturing world of the masculine workplace and the hostile environment of the modern city.

In an era of sensitive men exemplified by talk show host Phil Donohue and actor Alan Alda, the male-centered soap opera found an audience. While *Hill Street Blues* included significant supporting female characters, mainly there were men struggling to hold on to command, struggling to figure out their relationships with women who didn't understand their needs, traumatized by what they experienced in their jobs, struggling with addictions, and trying to negotiate the unspoken demands of male friendship. While also solving crimes and keeping the peace.

Bobby Hill and Andrew Renko, the young partners who almost died in the pilot, spend much of the series squabbling and enabling each other like a clichéd married couple. They are balanced by the more experienced partnership of the self-destructive JD LaRue and supremely cool Neal Washington. Captain Frank Furillo needs to constantly balance the demands of his job with the demands of the two women in his life, his ever-present ex-wife and his defense attorney girlfriend. By the end of the series, every male character—including the roll call sergeant who appears to have all of these issues behind him—is forced to face his inability to be what he believes a man should be. While also solving crimes and keeping the peace.

From the perspective of drama, the great challenge of traditional American television writing is that the main characters cannot change, allowing viewers to watch the show sporadically with confidence that they'll return to the same characters that they left several episodes back. In stories relying on an environment of unending threat of chaos and violence, the constancy of the show's stalwart Captain Frank Furillo is an even more necessary appeal of the show. But that is so for the rest of the cast as well, even those with flaws.

Familiarity breeds comfort.

The writing challenge is how to create a sense of the dramatically necessary change over time, transformation in characters and incidents, without rocking that central foundation of permanence that episode-based television relied upon.

There are traditional ways to create the necessary illusion. Continuing characters are revealed to be more human—that is, kinder, more sentimental, more empathizable as a series goes along. So the originally caricatured character of the undercover cop, Mick Belker, who alternates between raging fits and biting much bigger criminals is revealed as sweet, caring, and, most importantly, identifiable. Furillo's harridan of an ex-wife, Fay, is gradually revealed to be compassionate and more attractive than she initially appears. Howard Hunter, the comically fascist SWAT (special weapons and tactics) team leader has a tender side even if he does his best to conceal it. Each of the main characters—and there are many, with 14 actors identified in the opening credits of Season Two—has his or her characteristic flaws with a fuller backstory to be revealed, gradually changing the way viewers see them without actually requiring any change in the characters.

Furthermore, the large cast of characters allows for a regular shift of narrative focus, adding to the sense of change and development in the series, even if individual characters remain reliably unchanging.

With additional secondary—or "temporary"—characters appearing as victims, criminals, and complainants, and new lawyers, judges, and police regularly rotated in, *Hill Street Blues* is further able to disguise the lack of change in the running characters with temporary characters who can experience drastic change—up to and including death—and the continuing characters can offer the viewers catharsis through their responses to the change the secondary characters go through.

A New Look for a New Approach

Part of the mandate for *Hill Street Blues* was to help NBC hold on to an audience that was moving to the more "adult" and cinematic fare on cable. *Hill Street Blues* couldn't offer its audience the R-rated entertainment that cable could, but it could offer viewers a more realistic sort of story than they had previously seen on network. And the cinematic language, and to some extent the values, of Italian neorealism became part of the new realism of television drama.

Inspired by *The Police Tapes*, a 1977 documentary film about South Bronx police, the show's style expanded on the visual vocabulary and faux-documentary style of Italian neorealist filmmakers. The handheld camera and the wide and deep focus shots of Hill Street's station allowed viewers to watch action in the background while a scene played out in the foreground. The neorealists turned these camera techniques into a visual shorthand that meant "this is real." More subjectively, these deep focus shots made the argument that everyone is a part of their specific environment, the literal physical setting as well as the social environment of their community. Neorealist filmmaking stressed this relationship of character and environment as a part of a political philosophy that also served as a rich visual rebuttal to the star-centered film work of Hollywood at the time.

On television, the close-up was originally necessary to compensate for the small screen size but it also visually tells the viewer that whoever gets a close-up is more important than others. A close-upped hero—or villain—is set apart from their environment, one who acts alone and dominates whatever environment they inhabit. When the environment dominates the shot, as it did in *Hill Street Blues*—and has come to dominate in much longform television—it prepares the viewer for the likelihood that environment will dominate characters.

In Bochco's next big show, *L.A. Law* (1986–1994), this emphasis on environment expanded into the audio track, with background conversations regularly mixing in behind the lead characters and their stories. Improving television speakers made this technique possible, where traditionally background sound is muted so that important dialogue is clear. The added aural density in *L.A. Law* was used to depict a very different environment from *Hill Street Blues*. A corpse is rolled down a hallway and the viewers hear the bystanders' conversations—fragmentary comments about the deceased but also self-involved discussions ignoring the death in their midst. The audio reminds the audience that the show's main characters exist in a larger world full of self-involved and unsentimental people, grounding the show's more absurdly cynical tone.

Hill Street Blues also offered a more "realistic" set of characters than had been seen on television before. Headed by a police captain, Frank Furillo, who was a divorced, recovering alcoholic, the show depicted Furillo as more a reluctant bureaucrat than a cop, trying to manage budgets and political pressure, guiding his varied group of police officers through their jobs and personal crises, negotiating with neighborhood powerbrokers, exchanging a willingness to overlook minor crimes to get cooperation in protecting the innocent, trying to convince the bureaucrats downtown to do what was necessary rather than what was expedient, and still going to bed each night with a fabulously attractive defense attorney who spent much of her daytime creating more problems for him.

Each character on the show had similar problems to negotiate and compromises to make to get through their day. One seemingly small, but significant assault on television "realism" might illuminate just how completely the ground shifted with *Hill Street Blues*: The gangs with whom Furillo regularly negotiated peace treaties were all-black, all-Irish, all-Hispanic. They were not integrated. This may seem obvious today, but it was a controversial decision in the 1980s.

Comedian Freddie Prinze used to do a routine about being a Hungarican (Hungarian Puerto Rican) who was initiated into an all-black gang because they were bowing to public criticism that they were "not an equal opportunity gang."[4] Prinze was commenting on the counterintuitive tradition in U.S. television of depicting urban gangs as integrated so as not to depict any single ethnic group in a negative light in an episode. The practice is one of the more obvious examples of how American television serves to reinforce a socially normative message—nondiscrimination—by bending reality to an absurd level.

Hill Street Blues redefined that television reality. Bochco announced that *Hill Street Blues* would reinforce the social message of nondiscrimination, but in a more realistic way. The show would have authentically segregated gangs, and crimes would be committed by people of specific ethnic groups, without attempting to balance out the depiction of any group in a single episode.

In some episodes, black people were the villains, in some episodes Chinese, sometimes Puerto Ricans, sometimes Caucasians—although the Caucasians tended to the specifically ethnic as well: the Irish, the Italian, the Eastern European, the WASP (white Anglo-Saxon Protestant). Similarly, the heroes of the episodes would sometimes be black, sometimes Puerto Rican, and so on, without concern that the individual episode would end in an ethnically zero sum equation. Bochco argued that they would all balance out in the end and present a more realistic depiction of the world while also allowing for more dramatic possibilities.

This seemingly minor distinction was significant in its shift of the standard of realism but, more important to this discussion, it was a rejection of the idea that the values of the show had to be judged from a single freestanding episode. Bochco argued that what matters is the values in all the episodes taken together as a single work, not considered in isolation.

The idea that a show should not be judged on a single episode may seem obvious to some readers, but it wasn't in the 1980s and it actually still isn't. In 2011, David Simon found himself defending *Treme* (2010–) against criticism that its depiction of a rape victim was wrong because she didn't fight back in the same episode that she was raped, taking a few episodes to begin to recover before she struck back.[5]

The idea that every episode must balance its moral ledger is hard to escape in American television. But the move away from that single-episode standard frees writers to create much more challenging stories, and is an important move toward seeing a series as a single extended narrative work.

Complicating Television Stories in Longform

Each episode of *Hill Street Blues* represents one day, with the large ensemble cast of characters advancing several storylines in every episode. Some stories end in the same episode in which they begin. A longer story arc might end during any episode, but it was equally likely to continue for several more episodes. Viewers could not be sure which type of story they were watching, as both personal crisis and case stories could be wrapped up in the episode or be longer. Generally there were always a few personal stories slowly bubbling along under featured storylines.

Multiple storylines and arcs of different lengths make it possible, if not necessary, to tell more complicated stories than what had been the standard story structure of the stand-alone episode. One simple writing technique becomes much more exploitable in longform television: juxtaposition. Just as Shakespeare's plays tend to run variations on a theme through the different storylines—*The Merchant of Venice* juxtaposes a storyline of romantic love with another of love between parent and child, love between two friends, love of self, and love of money—multiple story arcs make it possible to juxtapose different characters confronting the same issue from different perspectives. Multi-episode arcs also make it easier to juxtapose variations on an issue with the same character over time, as well as between characters, so Furillo might confront a situation that demands he act on principle, only to be frustrated by someone doing the same thing an episode or two later, followed by another arc where his acting on principle leads to a catastrophic outcome.

This sort of juxtaposition allows for much more complicated and ambiguous moral lessons. Where a single episode might end up supporting or rejecting acting on principle, several different story arcs that explore the same theme in different ways might simply leave Furillo without a clear answer, and viewers disagreeing about whether Furillo is overly rigid or hypocritical or is just weakened by thinking too much.

This added complexity and ambiguity is generally seen as a more literary or "adult" sort of dramatic narrative, distinguishing it from the simple entertainment that American television has been expected to provide through most of its history. To consider it another way, with longform television, the simple short stories of *Magnum, P.I.* (1980–1988) and *Adam 12* (1968–1975) become the ambiguous novel of a season of *Hill Street Blues*.

The Inkling of Dramedy

Comedy in *Hill Street Blues* originally functioned at the level of the clowns in Shakespeare, a momentary break in the seriousness and perhaps a lampooning of one of the running themes of the show. Initially, viewers were presented with Mick Belker and Howard Hunter, the SWAT team commander as comic relief. Belker's rages and violence, his banter with the same tall black perp who seemed to appear in every episode, and the embarrassing phone calls he received from his mother. Howard, the proto-fascist head of the SWAT team (called E.A.T. here), had delusions of grandeur, feeling he should have Furillo's job, wanting to treat all urban problems as military challenges.

As the series develops, Belker and Howard become more rounded and the comedy moves to secondary storylines about absurd cases and one-off characters, approaching the comedy of the sitcom *Barney Miller* (1974–1982) from the other direction. In both *Hill Street Blues* and *Barney Miller* the absurdly comical character could become poignant at any time, blending drama into the comedy.

What both series share is one of the deep roots of dramedy, a dissatisfaction with the limits of the genres and narrative formulas that had developed on American television. Just as adding long-arc stories and personal lives of characters to a cop drama expanded the dramatic possibilities, adding comedy to the serious drama (or serious moments to a situation comedy) significantly expanded the boundaries of stories that could be told as well as the attitude the show could take toward those stories.

Besides offering two different emotional experiences in one show, the blend of comedy and drama also became an important tactic in the expansion of network television into a longform narrative. As *Hill Street Blues* showed, a series could present multiple story arcs, some the traditional length of an episode-long story, others spanning several episodes and some a whole season. This means that the main dramatic story might not resolve, or might resolve in a tragic manner. But the injection of a comic arc in each episode, one that pays off at the end of the show, say a series of competing pranks or a running gag or a misunderstanding finally revealed, can provide a simple sense of closure—and feel-good closure, at that—at the end of an episode in which the major story and issues remain unresolved.

Hill Street Blues rings interesting variations on this blend, by sometimes revealing an absurd character to be bittersweet or even tragic. At other times, an absurd character or plotline is just relief, like the dance number in a Bollywood film. As with the varying lengths of story arcs, the range of different possibilities for resolution to comic arcs deepens the viewer experience by making the series less predictable.

Present-day viewers may assume this is the way television has always been but this American convention of the hour-long drama with comic relief and the situation comedy with a deeper poignant message developed as American television writers began to think of the show in terms of multi-episode storytelling, rather than the single stand-alone episode.

A decade into the new century, we expect some comedy in our drama and some drama in our comedy. It is the rare show that is simply serious, for example *In Treatment* (2008–), *The Killing* (2011–), *Boss* (2011–), and *Rubicon* (2010). A couple decades after American television began blending comedy with its drama, many viewers—and some critics—interpret a show that takes itself completely seriously as simply bad television.

Adjusting the Mix: *L.A. Law*

By contrast, *L.A. Law* combined even more soap opera elements—and more women, allowing more romantic entanglements—with a legal drama, but added a cynical humor absent from *Hill Street Blues*. Upon finding a partner dead in his office after a long weekend, Arnold Becker's first act is to claim the better office for himself. Rolling the body out of the office, the body slips and falls face-first in the carpet, with a floor-level camera shot. The managing partner tries to marry the firm's new Hispanic lawyer to his housekeeper because she needs a Green Card. A hot receptionist who is the object of lust for at least one young lawyer reveals himself to be a pre-op transsexual who met the (married) dead partner at a gay bar, with much amusement taken in the discomfort of the lusting young lawyer and the despair of the dead partner's previously clueless widow.

Here, the humor is less a matter of dramedy, than a stylized sense of cynicism as realism that permeates even the serious scenes. An Hispanic lawyer is hired as much for his potential as ethnic window dressing as for his legal skills. A black attorney quits the firm because he believes there's only room for one minority to become partner, and he's probably right. Sexism hounds the women in the office with sexist clients, double standards for female lawyers, and Becker, the divorce lawyer, going unpunished for his shameless womanizing of associates, clients, and any other woman in between.

Besides women and cynicism, *L.A. Law* also added sex to the *Hill Street Blues* formula. Sexuality with a smile, or at least a smirk, as the show pushed against network restrictions for a more "realistic" depiction of sexual relationships, sexual relations, and how people talked about them. The series avoided using banned words by describing them, "a sexual position normally described by two numbers" or a woman telling a flirtatious man "there's a two-word description for what I'd be if I let you to kiss me, Mr. Kusack, and the second word rhymes with 'freezer'" as the show fades to black with the man—and the audience—working his way through rhymes toward the two words "cock teaser."[6]

Between its cynical attitudes and titillating humor, *L.A. Law* could offer a challenge to the "adult" appeal of cable entertainment while remaining a network show. In fact, the humor of the show was largely a tactic for getting into subjects and sexuality that network censors would not allow to be directly addressed. This prophylactic use of comedy in drama initially shaped, and then was shaped by David E. Kelley, a Boston lawyer Bochco hired to write for the show. Kelley took over *L.A. Law* a few years later and then left in 1991 to create his own longform television shows, including *Ally McBeal* (1997–2002) and *Boston Legal* (2004–2008), both building on the *L.A. Law* formula with larger doses of absurd and sexual humor.

Self-Exploration through Doogie Howser

When Bochco left *L.A. Law* he had his pick of industry jobs and decided on a deal where he would produce 10 ABC series for Fox Productions that Bochco could sell elsewhere if ABC turned any of them down. This was a rare level of freedom for a creator of American television and it led to three reconfigurations of what network television could be: *Doogie Howser, M.D.* (1989–1993), *Murder One* (1995–1997), and *Cop Rock* (1990).

Doogie Howser, M.D. was a medical drama crossed with a coming-of-age sitcom, as it featured a 16-year-old child prodigy doctor. It may be the first show introduced to the public as a dramedy, the label appearing in a September 1989 *LA Times* interview with Bochco, which claimed the term was created in 1988.[7] The show is more drama than comedy, but is only a half-hour long. The humor is more smile-inviting than laugh-inducing, like the contemporary *The Wonder Years* (1988–1993) or *Leave It to Beaver* (1957–1963) of earlier years.

Beyond another unexpected blending of genres, the show is significant because it is evidence of the way that Bochco makes television a personal activity. Bochco's father was a central figure in his creative life, to the extent that Bochco's end-of-show logo was an animation of his child prodigy father, as an adult, playing the violin. *Doogie Howser, M.D.* became a gentle meditation on his father's life, and on the relationship between a father and a son, in the same way that Dennis Potter's work frequently examined fictionalized versions of his life, his suffering, his relationships with women, and his relationship with his past.

Central to the series was Doogie's relationship with a strong admirable father, a fictional relationship that is hard to avoid reading as Bochco's variations on his relationship with his father. Bochco's father was in his fifties when he was a child, so the series gave him a change to access his father's earlier life by creatively projecting himself into his father's childhood. The series was canceled after four seasons, at a time when Bochco was about to have his child/father-surrogate begin to live more of Bochco's life. In the fifth season, Doogie was to have turned his back on medicine and set out to become a writer in a single story spanning the entire season.

While Bochco opened doors to the creation of dramedy, his hour-long blending of drama and absurd comedy was quickly dominated by David E. Kelley. His half-hour series *Doogie Howser, M.D.*—perhaps because dramedy is not seen as an appropriate narrative form for children—ended up opening a door to dramedy that no one seems to have gone through.

Murder One: The Single-Story Season

Roone Arledge of ABC Sports watched his television on November 24, 1963, as news commentators showed film of Jack Ruby killing Lee Harvey Oswald in the Dallas Police parking garage, backing it up and running it forward over and over again to explain what exactly had happened.

It gave Arledge an idea and soon ABC Sports, then all of television sports, showed viewers important plays in slow motion instant replay.

Steven Bochco watched the months of all-day televised proceedings of the OJ Simpson murder trial in Los Angeles that ran for most of 1995 and he had an idea.

The result was *Murder One*, a legal drama that reflected the reality of the trial that millions of Americans had been watching. It would follow a single murder trial through 23 episodes.

Arledge's idea changed television sports coverage forever by replaying every moderately interesting play and making instant replay, not the on-field official, the final arbiters of close calls. *Murder One* lasted for two seasons, advancing the idea of a series telling a single season-long story. Its greatest influence, though, is seen on cable television, most completely in the television series *Damages*.

The story is deceptively simple. A bad boy movie star is accused of murder and comes to the best defense attorney in Los Angeles. The defense attorney works with his lawyers and investigators to free his client and find out what actually happened.

This plotline could be a one-off episode of a legal drama, but it is instead the starting point of a 23-chapter legal procedural and mystery, with broader themes explored along the way. The length of the single storyline allowed for a different type of storytelling. The show's methodical pacing is reinforced by the central figure of Theodore Hoffman, the still, moral intelligence at the heart of the series.

The other dominant character is Richard Cross, a shady *macher* in Los Angeles who was the lover of the older sister of the murder victim who has died in one of his apartments. Cross enters the series as an interested party who knew the key players in the murder trial of a young movie star, but the series comes to reveal Cross to be the true antagonist to Hoffman, a force behind the scenes who seems to be causing witnesses to recant or disappear, evidence to appear or disappear, and political pressure to circumvent investigations and change charges. He knows more than he is telling and, even while the murder trial continues, investigating Richard Cross becomes the main focus of the series.

That is the simple plot, but captures nothing of the mood and the themes.

The Simpson trial had taught the television audience that actual lawyering is slow and fixated on detail and *Murder One* made the most of it. Where the more recent and successful serial show *24* keeps the adrenaline pumping, *Murder One* establishes a methodical pacing in which shocking events and revelations share the same rhythmic pacing as meticulous legal research, reinforcing the idea that the viewer was in this for the long run. A murdered witness is revealed within the same rhythm of inevitable advance as the discovery of a mistake in an arrest warrant, which is the same emphasis as is given to an associate pushing too hard to get second chair, or Hoffman saying he won't be home for dinner. A search for a hidden camera in an apartment is four minutes of minimal vocalizations and shots of men running their hands along surfaces in the dark; the eventual jury verdict includes the reading of the entire formal language, just as the audience had seen during the OJ Simpson trial. The verdict is presented with no more hype than several other points in the episode, even though Hoffman loses the case. Instead, the lawyers immediately, methodically move into the next phases of the trial, the appeal, and management of public perceptions.

The surprise is that it is still suspenseful, perhaps more suspenseful than most network television because the rules of this new approach to the legal drama are not clear. *Murder One* is confident that it has the time to tell its story. The show moves at a pace more familiar from daytime soap operas, with talking heads seriously discussing issues of law and strategy rather than intrigues of the heart, enhanced by harpsichord cues that invoke the musical accents of the classic daytime dramas. Alternately, the pacing of the season-long story might be better compared to the contemporary best-selling lengthy legal thrillers of Scott Turow.

The cumulative effect is powerful, throwing the focus to Hoffman at a formal level, as the story's rhythms echo his understated methodical approach to everything. Ironically, the show's avoidance of traditional dramatic pacing makes the slow ratcheting up of the tension more compelling. Theodore Hoffman, played by the perpetually *sotto voce* Daniel Benzali, is a large bald man who gives away nothing and is all the more threatening for it. He seemingly has contempt for the corrupt world of celebrity and excess in which he thrives and the season reveals his increasing doubts about the cost of his work, leading to his withdrawal from his profession. The central conceit of the show is that this defense attorney—an occupation generally portrayed as the moral equivalent of a drug-dealing prostitute—is, in fact, the most moral force in a corrupt world.

The stillness and understatement of Hoffman is juxtaposed with a more evasive Richard Cross, played by Stanley Tucci. Cross is flamboyant and charismatic from the start, but is gradually revealed to be a man of great corruption, a tempter, and a destroyer who may be the real killer who has framed Hoffman's client while appearing to help him.

That description doesn't capture what the series does with him because, while the trial plods methodically along, the Richard Cross material—and the Richard Cross and Theodore Hoffman interactions—move into a gothically romantic vocabulary, with Cross becoming a Satanic figure in contrast to Hoffman's rock of faith. As a matter of writing and performance, it is a significant accomplishment to advance the realistic trial material while the heart of the show goes off into a florid gothic story without either element suffering.

Those who want to see Dennis Potter's most direct influence on Bochco's work should consider the relationship of *Murder One* to Potter's long-suppressed television play *Brimstone & Treacle*, filmed in 1976 but not broadcast until 1987. (In the meantime, the script was made into a 1982 film starring Sting.) In what is sometimes called a Pinteresque plot device, a strange young man enters a middle-class home to become the caretaker of comatose young woman. In the course of the story, he is kind and he is sadistic; he may be a demon, or an angel. In the end, he attempts to rape the comatose woman but she awakens, cured, and the young man runs away. To oversimplify a disturbingly ambiguous meditation on good and evil, viewers are forced to consider how his evil act has cured her.

Murder One copied nothing from *Brimstone & Treacle*, but in Cross and Hoffman, the series provides Bochco the opportunity to explore issues of evil and good in his own way. As the season nears its end, Cross becomes more insane, or manipulative, or evil. Viewers learn of Cross manipulation of corrupt politicians and innocents, his perversions, and his conspiracy with a corrupt psychiatrist. Cross seems to be revealed as an all-powerful megalomaniac as his motives remain unclear and gradually seem to transcend reason. A climatic confrontation with Hoffman builds to a point where Cross could easily be revealed as a Satanic force.

But the story does not end there. There are several more hours to come and in those remaining chapters viewers receive explanations and reversal. Cross is revealed to have had AIDS for years and, while he has disguised its effects with steroids, he has been slipping into dementia caused by the disease.

The reversal is far more shocking, though. The obvious villain Cross is, in fact, an increasingly powerless man who did not kill the woman. She was killed by a South American mobster that Cross had been in business with, who has forced Cross to cover for him. (Re-watching the series reveals carefully planted evidence that supports this.) In the end, Hoffman's wrongly convicted client is freed only by the villain's unexplained act of kindness. The viewer is left to decide whether Cross exonerates him as an act of penance, as a way to torment the contemptible District Attorney, because the woman he loves asks him to do it, or simply because he is now so weak that he cannot avoid doing what he is told.

Whatever the interpretation, the first season of *Murder One* ends in moral ambiguity, nobody to punish, and with plot twists network television viewing had trained the audience not to expect. Bochco was able to explore the broad moral issues while remaining within the demands of genre television.

Bochco had felt that Americans who watched hours of daily sidebars and boring testimony in the OJ Simpson trial would be ready for a similar fictional experience, including an ambiguous ending.

It was a reasonable bet but wrong.

Murder One confirmed the limits of network television in delivering longform television. The 23-hour series told a complicated story that essentially required the viewer to see every episode. Even with ever-lengthening summaries of plot points from previous episodes, if viewers missed an episode, they had a lost a chapter of a story of shifting allegiances, emerging—and possibly misunderstood—evidence, subtle maneuvering, and tonal shifts in performance that suggested what was actually happening.

ABC programmed the series against top-rated NBC shows *ER* and then *Seinfeld*, with preemption and reschedulings that made it difficult for viewers to find when it was telecast. In the days before TiVo, OnDemand, and Hulu, as well as no Internet blogs to provide detailed plot summaries of the missed episodes, it was pretty much guaranteed that no viewer would be able to see the entire series.

Murder One in 1995 stretched network storytelling beyond its ability to be seen. In practical terms, it was nontelevision narrative without the knowledge that a DVD set could eventually rescue it from oblivion. The series did find critical acclaim in the first season and the network renewed it for a second season split to follow three cases instead of only one, but the ratings did not improve and the show was canceled.

Not surprisingly, the series found a second life when the two seasons were released a decade later as DVD sets, allowing most viewers their first opportunity to see the entire story as it was meant to be seen. For the first time, viewers could see the gradual ratcheting up of tension and the subtle changes that are so important to this moody series as it gradually turns from a simple murder trial to a metaphorical battle between an everyman and a force of evil in human form, only to reveal the monster to be a fragile, dying man.

Twelve years later, *Damages* (2007–) premiered on FX.

It takes nothing away from *Damages* to say that the series is a simpler, shorter variation on *Murder One*, excepting a very important shift from a basic male melodrama to a heightened return to women's melodrama, though still in the now conventional workplace setting of the male soap opera. Supposedly, the show originated in a desire to explore a female mentorship relationship and only came to lawyers later in the process. Glenn Close becomes the amoral mentor to Rose Byrne's Ellen Parsons, at one point in the first season attempting to have her killed, but also supporting her career because she sees a version of herself in the younger lawyer.

Unlike *Murder One, Damages* exists in the cable world of instant reruns, streaming, and OnDemand video, and a DVD set waiting at the end of the season. It also works with the now standard 12-episode cable season, increasing the chances that the audience will see the whole series as it airs.

Where television has often given viewers the "father-son" narrative, or an older man–younger woman mentorship like the one that becomes a marriage in David E. Kelley's *The Practice* (1997–2004), *Damages* reinvigorated the form by adding a "mother-daughter" mentorship model with a bad, though somewhat justified, mother-figure in Close's Patty Hewes. The series offers a story of two strong women—the older mentor who made her way to the top of a man's world but continues to struggle as an outsider; and the other a young woman who wants what Hewes has, who is tough enough, though not yet experienced enough, to get it. The significant change from standard television is that the two lead characters are both involved in a clearly self-destructive relationship, with a theme song—"When I Am Through with You (There Won't Be Anything Left)" by VLA—ambiguously promising that one or both of the women will be destroyed at the hands of the other.

Given that, it is a bit surprising that neither Patty Hewes nor Ellen Parsons change that much over the run of the series, although each season begins with a resetting of plot details about where Ellen is working and the condition of Patty's home life. Once the pieces are set for a new season, though, they are the same characters they ever were, with Parsons being a more underdeveloped version of Hewes. After the first season, Ellen knew that Patty had tried to have her killed and the second season started out with her seeking revenge, only to end up working with Patty again. Then Ellen went to work for the District Attorney, but ended up working for

Patty again. Then Ellen was working independently, and ended up working with Patty again, ending with her vowing revenge when Patty sabotaged a deal, nearly getting Ellen and her client killed. The next season will begin with the fallout of Ellen helping Patty's son take his child away from Patty.

As with *Murder One*, the series is based on a connection to the viewers' understanding of events in the real world. The season-length legal arcs of *Damages* are cases of corporate corruption being pursued by Hewes & Associates, using the "ripped from the headlines" approach mined by *Law & Order*. Hewes goes after an Enron-ish case, a Madoff-ish case, a case of energy company pollution and corruption that might have reflected cultural negativity toward BP, and a Blackwater-ish case. At the same time, equal weight is given to the failure of Patty's interpersonal relationships. (It may be worthy of note that a highly praised series about two strong women presents them both confirming the cliché that a woman cannot have a successful career and a successful home life, that the pursuit of a career turns a woman into a cross between a gothic villain and Lady Macbeth.) Eschewing the methodical pacing of *Murder One, Damages* hews closer to the adrenalized pacing of *24*, heightened by episodes beginning in the past or the future, to generate mystery and add tension as the episodes rush to fill in details and move us toward explanations.

Damages offers its viewers the opportunity to see corrupt corporate scofflaws revealed as unambiguously evil and then punished in a way that they never are in real life. In that arena, the show functions mainly as an engine to engage viewers' desire for revenge. That element was missing from *Murder One*, which made a point of frustrating our moral certainties and desire for revenge. At the end of the first season, there was no one to punish. *Damages* frustrates viewers' expectations in terms of plot points, not judgments, and the surprise twists have become a part of the formula.

Cop Rock: Musical Narrative Television

Cop Rock may have been the blend too far.

The Bochco series became synonymous with creative overreaching, an easy punchline as a shorthand for television failure equivalent to film's *Heaven's Gate* (1980).

While Bochco had built a career on successfully blending different television genres to reinvigorate them, with *Cop Rock* Bochco attempted to blend the television cop show with a nontelevision form—musical theater. Since *Cop Rock*, there have been regular attempts to create a successful form of musical drama for television, culminating with the success of *Glee*. The attempts along the way suggest where Bochco may have gone wrong.

One established "fact" about *Cop Rock* that is due for reevaluation is that it was inspired by Dennis Potter's *The Singing Detective*. Bochco did point to *The Singing Detective* as *an* influence, but the specific statements were more ambiguous. In 1992, after *Cop Rock* had failed, Bochco said in an interesting, bantering Q&A with Potter at the Museum of Television and Radio in New York, that "we just stole you clean."[8]

The series did no such thing, so it's hard not to read that as straight-up irony. *Cop Rock* is pure Broadway musical in its conventions, not the modernist deconstruction of narrative with the social meaning of music explored in hallucinatory sequences of ironic lip-synching artifice of *The Singing Detective*.

The reason they are so different is, as Bochco acknowledged in the same appearance, American television is "not fundamentally an art medium."[9] Potter had self-consciously made a work of deconstructive, modernist art, while Bochco's *Cop Rock* used the conventions of popular musical theater to broaden the vocabulary of an entertainment medium. Among the reasons Bochco might have credited Potter's show is that he wanted to downplay the fact that, before beginning the series, he had been approached by people who wanted him to consider adapting *Hill Street Blues* into a musical for Broadway, although Bochco has said he was already thinking about doing musical television.

Regardless of origin, *Cop Rock* is actually a good adaptation of the rules of the musical narrative to the television screen, a show with the potential to become successful if it had had the time. The basic story was an L.A.-based *Hill Street Blues*-type show with a large ensemble, one major dramatic long-arc, lesser multi-episode arcs, and single episode closure. The only alteration from *Hill Street Blues*, admittedly a significant one, was that the show followed the basic convention of musical theater that characters break into song and sometimes dance in a more fantastic space, to reveal their inner feelings before returning to the dramatic events.

In episodic terms, *Cop Rock* also had the potential to attract audiences to a weekly performance of popular music by a single composer, as in the case of the pilot that featured songs by Randy Newman. Commercially, a network alternative to the televised music videos of MTV would have been worth considering.

If one watches the series as a DVD novel, immersed in multiple episodes in a sitting, the weird logic of the musical becomes the norm and it becomes a serious cop drama where the emotional lives of the characters are revealed, not in soap-ish conversations, but in songs. (Unfortunately, this is not an easy experience to achieve at the moment; the series is only available to those willing to use online commerce to purchase low-quality bootleg collections of out-of-order episodes.)

Even if the series spent the last 20 years as an object of ridicule, *Cop Rock* kept the idea of a television musical drama a subject of discussion, and the variations that have followed reward a bit of examination. In retrospect, Bochco made several mistakes in judging the limits of acceptance in the American audience.

In a general sense, there was a decent chance that the cop genre could be turned into musical. Rogers & Hammerstein's 1943 *Oklahoma* had become a highly successful play and Hollywood film even though it was a serious cowboy romance balanced with bleak comedy and the very un-"popular" art form of ballet. Unfortunately for Bochco, he was adapting musical theater at the end of the 1980s, when Broadway was more cut off from mass entertainment than it was in the time of *Oklahoma*, or even today when Hollywood films are regularly being turned into musical theater.

At the time of *Cop Rock*, the tides were literally shifting from the Broadway theater as an origin of film and television entertainment to a place where film and television ended up. Within a few years, Disney films were being turned into the Broadway musicals we take for granted today, including *Beauty and the Beast, Little Mermaid, The Lion King, Tarzan,* and *Mary Poppins.* The trend continued with *The Wiz, Sweet Charity, Chitty Chitty Bang Bang, the Producers, Hairspray, Big, Footloose, Victor/Victoria,* and *The Addams Family,* to name only the bigger hits. Making a *Hill Street Blues* Broadway musical might have been a more successful idea than *Cop Rock*, adding television to the wave of recycling film into theater that was just beginning in earnest.

At the same time as *Cop Rock* was failing on ABC, Larry Gelbart's *City of Angels*, the unacknowledged adaptation of the themes of Potter's *The Singing Detective*, was a critical and commercially successful musical on Broadway. *City of Angels* is a musical comedy about a mystery writer struggling against the Philip Marlowe-esque detective he created as his novel is being adapted into a Hollywood film, where the clichés of film noir invade or shape his perception of his life. Very successful on Broadway, the play never made it to any video media.

A simple cultural explanation for the failure of *Cop Rock* is that American audiences want their television to be "realistic." Even while shows can, in fact, be highly stylized and artificial, their artifice must follow established conventions. As long as they do, vampires and demons and leech monsters and aliens—from *ALF* (1986–1990) to *X-Files* (1993–2002) to *True Blood* (2008–)—can be accepted by American television audiences as long as they are "realistic" within a television sense of the term. The puppet alien on *ALF* is "realistic" because he acts like sitcom characters have always acted, in the set and setting of a recognizable middle-American sitcom reality. *True Blood* can present shape shifters and vampires and angry witch-ghosts because Sookie, the lead character of the show, is just as disgusted as the audience by the absurd plot twist revelation that she is a fairy.

Television realism—in fiction and nonfiction—is about conforming to accepted conventions. Morally challenged cops breaking into song and dance to express their unrequited love on the way to a crime scene is not "realistic" within those conventions.

As discussed in Chapter 2, *The Singing Detective* influenced the use of music in American television but the most commercially successful uses of singing in narrative television have used characters that could realistically be expected to sing. The musical number is rationalized. The characters on *I Love Lucy* (1951–1957) or *The Dick van Dyke Show* (1961–1966) can put on a show and sing and dance because they are performers. *Ally McBeal* gives viewers wacky lawyers who like to take over a bar and sing. Most successfully, *Glee* (2009–) presents a group of singing high school students who regularly break into song.

A related explanation for the failure of *Cop Rock* is that, musically, U.S. television audiences want what they've already heard. *Cop Rock*'s emphasis on all-new songs for every episode meant that viewers had to decide they liked the music based on only a single hearing, without the ability to listen again, where American music consumption is based on heavy rotation of recognizable songs with new music gradually introduced within a context of familiarity.

Songs on American television at the end of the 20th century are essentially the re-presentations of the familiar, whether simply as soundtrack music or as songs performed by characters within a show. For better and worse, long-arc shows that used musical performance by characters, from *Ally McBeal*, to *Eli Stone* and *Viva Laughlin* (2007) to *Glee*, have used recognizable hit songs, generating comfort through familiarity, but also tapping an existing emotional response that is reattached to the television show. To hold an audience, the musical on American television cannot be a cohesive original work, it needs to be a collage of existing songs and cultural references. This creative technique might well have been developed in further interesting ways if the cost of licensing songs for American broadcast hadn't skyrocketed in the 1990s, making it difficult for television shows to afford to use more than one or two songs in an hour.

Songs on soundtracks, rather than performed by characters, seem to have derived from these conventions and restrictions. The convention of one meaningful (and cheaper) indie pop song over a montage of characters in separated scenes to encapsulate the emotional subtext of the episode is so widely and deeply engrained in television drama that it can be mocked in the early seasons of *Supernatural* (2005–), the series about demon-slaying brothers. Each time the car stereo played a pop song it was an inescapably clear commentary on what the brothers were doing.

The convention has settled into the use of a different (mildly) popular song each week as the show moves into the end credits, then continuing over them. *Northern Exposure* (1990–1995) and *Gilmore Girls* (2000–2007) are two prime examples of this in earlier network shows, but most one-hour HBO dramas have are firmly committed to the use of a period pop song as a summary statement on the show. *The Sopranos* (1999–2007), *Carnivale* (2003–2005), *Boardwalk Empire* (2010–), and *Treme* (2010–) are just a few examples.

In this usage of music, the logic remains that of Dennis Potter's use of the pop songs in *The Singing Detective* and *Pennies from Heaven*, albeit in a "familiarity breeds inattention" sort of devolution. The song serves as an ambiguous closing statement on the events in the episode, providing a new context and final comment on the episode while, at the same time, often suggesting a new meaning to the lyrics of the song, based on their juxtaposition with the new narrative and emotions that the episode offered.

With the exception of *Treme*, however, the shows do not make the music the central element that Potter did in his most famous works. Potter wanted to explore the popular song as a sort of sacred music of a secular age, through which people could feel a sense of resolution toward the unresolvable difficulties of their lives. *Treme* explores how music is the heart of a community, not the specific use of specific songs, but the simple acts of creating and consuming music. The use of Bob Dylan's "All Along the Watchtower" in the final story arc of *Battlestar Galactica* (2004–2009) may come the closest to Potter's strategy of using a familiar song to generate feelings of familiarity and alienation in the same moment, with a sort of cosmic lip-syching, suggesting the song is literally a divine text.

Northern Exposure, WKRP in Cincinnati (1978–1982), and *Miami Vice* (1984–1990) each reflected the symbolic form and realities of their earlier decades' radio-based music, with mixes of popular songs literally broadcast over the shows' radio stations to provide a commentary or set the reality of events. *Miami Vice* was less literally grounded in a specified radio station, but its iconic pilot moment is of Crockett and Tubbs getting into their car and Phil Collins's "Something in the Air Tonight" playing as they drive across Miami to a final battle. The soundtrack of *Miami Vice* was very much an FM album rock sampler of its day and initiated the "rock radio hit as soundtrack to life" convention that flows down to the indie hit song montages of countless television dramas to this day. *Glee* succeeds by reflecting the new century's reality of music as a location of competition for life-changing transformation through successful performances of authenticity.

The dominant meaning of music on television in the new century has been the *American Idol*–type singing reality competition shows. *Glee* is only an extension of the fictionalization—or narratization—that takes place on a show like *American Idol*, where a set of cover performances of popular songs are interspersed with melodramatic (literally—with melody and drama) depictions of the personal struggles of the singers, with each young singer having a familial, interpersonal, or personal issue that inspires them or must be defeated through song. *High School Musical* (2006), the unexpectedly successful Disney television movie and its sequels, may have brought the Broadway musical formula to a new generation of television viewers but *Glee* found the way to turn the dominant reality of television music—young people singing covers of pop songs as a transformational struggle—into a teen soap opera with multi-episode arcs that satisfied the network need for freestanding episodes by filling each episode with a set of thematically-related song performances that end up teaching a valuable lesson.

Which leaves the one last exception by which American television shows can become musicals—the very special musical episode. *Buffy the Vampire Slayer* (1997–2003), *Scrubs* (2001–2010), *Chicago Hope* (1994–2000), and *The Drew Carey Show* (1995–2004), all established shows with a loyal audience, have been allowed to do a musical episode filled with original songs. There is an explanation for the singing involving a mental disorder or a dream sequence or a curse and all are presented within an ironic sensibility that tells the audience "we aren't taking this seriously, either," but it is essentially the postmodern version of *The Dick Van Dyke Show* or *I Love Lucy* doing an episode where everyone performed at a party. The increased ability to rewatch these episodes through DVD sets, YouTube, and streaming video have also removed that problem that plagued *Cop Rock*. In repeated viewing, some of the songs have become popular with their fan bases and *Buffy*'s musical, "Once More with Feeling," has become one of the most popular episodes of the series.

In 2000, David E. Kelley, Bochco's former collaborator, used the final episode of the third season of *Ally McBeal* to show how *Cop Rock* might have been successful. "Ally McBeal: The Musical, Almost" begins in *Cop Rock* territory, repeating that pilot's use of Randy Newman music in a Broadway musical form, except using only his established hits, not new music. Having established the episode as a musical using familiar songs, Kelley ups the ante. The anxiety of a double date with her parents sends Ally into a Dennis Potter hallucination where everyone is suddenly conversing by singing fragments of old standards, and Ally's younger self shows up to sing her judgment of her father.

It's an interesting response to *Cop Rock*, made even easier to accept on network television by the show being comic, rather than a serious cop show. The episode of *Ally McBeal* demonstrates just how formally conservative *Cop Rock* actually was.

Finally, the very simplest explanation for the failure of *Cop Rock* has nothing to do with music. It is that the series cost too much. Just as high production costs eventually led to Bochco being removed from *Hill Street Blues*, ABC would probably have been willing to leave *Cop Rock* on the air for another season to build an audience if Bochco could have kept the costs down. Creating another large ensemble cast show, with the added cost of creating and recording a whole show's worth of original songs each week, meant that *Cop Rock* needed to be a hit from the beginning.

Given how central the *Hill Street Blues* model of television narrative has become, it's worth remembering how close the show came to being canceled in its first season. The show survived to become a major hit because NBC executives saw that it tested well with the Midwestern urban demographic who, before the threat of cable, could be comfortably ignored. The concerns about disappearing audiences created a new television environment where *Hill Street Blues* could succeed.

The result was that people learned to write and watch television in a different way, but it was not due to any conscious decisions to make a new type of television. Bochco wanted to tell stories about too many characters to fit into a single episode. Blending elements of soap operas provided a model for doing it. Network anxiety about attracting new demographics bought the show enough time that audiences developed a taste for that sort of storytelling.

L.A. Law built on the *Hill Street Blues* formula to create another long-form series that could satisfy the episodic needs of network, while also training the audience to accept a more stylized form of narrative, saturated with a cynical and absurd humor, sexiness, and an entertaining presentation of social issues.

Doogie Howser, M.D. moved the blending of genres further down the road while also showing there was room for a deeply personal meditation within a prime-time network television show.

Murder One revealed the limits of the long-arc drama on network television, even as it later proved a success when reissued as a single season DVD set. But it ended up setting the model that would become much more successful on cable in *Damages*.

And *Cop Rock* failed in such an interesting way that it demonstrated the limits of the network television audience. Even as Bochco showed how to blend genres on television, he could not succeed in getting American audiences to accept the "realism" of the Broadway musical in a serious dramatic show. By contrast, the successful *Smash* (2012–) is a show about the making of a Broadway musical that rationalizing all musical performances within the "realism" of traditional dramatic network television.

What remains characteristic of all of Bochco's shows, including the ones that he continues to make today, are that whether on network or cable, Bochco continues to work from the templates of network television shows that he learned starting out at Universal. In each show, he respects the basic assumptions and then attempts to change what a show can do within the demands of those conventions.

The long-term effect was that Bochco's shows brought the continuity of the soap operas into other genres by developing the blend of multi-episode arcs with and episodic closure, satisfying the need for the first time viewer while also acknowledging a continuing audience.

This is the difference in thinking about what sort of story a television series tells. In literary terms, Bochco's shows began to change the way American viewers and creators thought about television episodes, from independent short stories to chapters in a novel.

Chapter Four

Stealing the Soap and Longform Network Television

In Which We Consider Further Attempts to Create Longform Dramatic Television Shows, from Modified Anthology Shows to Soap-Ish Multi-Episode Arcs and the Resulting Anti-Heroes

You're going to have a self-contained episode every week, just like opening a book and reading a fairy tale, and we're going to solve a crime each week, but for those loyal enough to watch every week, there'll be all kinds of hidden messages and there will be arcs for all the characters and terrible things are going to happen to many of them.

—David Greenwalt, promoting the new series *Grimm* at Comic-Con 2011[1]

These days it is taken for granted that a dramatic series has a long story of some kind, but that hasn't always been the case. With *Bonanza* (1959–1973), *Dragnet* (1951–1959), *Mannix* (1967–1975), and even *Rockford Files* (1974–1980), episodes just happened. When the men from *U.N.C.L.E.* finished a case, someone hit the reset button and it started all over again next week.

Starsky and Hutch (1975–1979) did not have a long arc.

Miami Vice (1984–1990) had some multi-episode arcs, but the episodes stood alone except for very special linked episodes like the ones where music star Sheena Easton played the true love of Sonny Crockett's life (Don Johnson). It is hardly a spoiler to say that she was killed at the end, since it would have been the same on *The Mod Squad* (1968–1973), *Ironsides* (1967–1975), *Knight Rider* (1982–1986), or *Baywatch* (1989–2001). One of the conventions necessitated by episode-centered storytelling is that no new character of significance in the life of the lead characters can survive for more than a couple episodes.

The Long Arc and Anthology Series

Roy Huggins was a novelist who moved into film writing in the late 1940s, then to television production at Warner Brothers in 1955. After a brief stint at 20th Century, he settled in at Universal for 18 years. His short stories and a novel were turned into three series: *77 Sunset Strip* (1958–1964), *City of Angels* (1976), and *The Rockford Files* (1974–1980). Other series he created and produced include *Maverick* (1957–1962), *The Virginian* (1962–1971), and *Baretta* (1975–1978).

Most important for this discussion, Huggins unintentionally created an early template for longform American television with his show *The Fugitive* (1963–1967). *The Fugitive* was one of many anthology series, where a character moves into a new location for an episode, gets involved in a story, and then moves on at the end. A modern variation on the wandering cowboy hero, the formula is seen is shows as diverse as *Route 66* (1960–1964), *Then Came Bronson* (1969–1970), and *The Incredible Hulk* (1978–1982). This form was already a narrative enhancement of the pure anthology shows such as *The Twilight Zone* (1959–1964) or *Alfred Hitchcock Presents* (1955–1962), where a host's introduction each week is the only continuity from episode to episode.

The Fugitive's innovation was that it had a continuing, series-long story. *The Fugitive* featured Dr. Richard Kimble, an Indiana doctor falsely accused of the murder of his wife when the real killer was a one-armed man. Huggins always insisted his inspiration was actually found in Victor Hugo's novel *Les Miserables*, where Jean Valjean, who stole a loaf of bread, is pursued for decades by an obsessed Inspector Javert. Many, however, see inspiration in the notorious 1954 trial of Dr. Sam Sheppard of Ohio for the murder of his pregnant wife, the OJ Simpson trial of its day. Sheppard claimed a bushy-haired man killed her and knocked him out, but he was convicted. (After years of appeals, he got a new trial in 1966 and was acquitted.)

The long arc of *The Fugitive* is that Kimble escapes custody after being convicted and is on the run while seeking the one-armed man. He is pursued by police lieutenant Philip Gerard, who has made it his mission to catch Kimble. This long arc hangs over every episode, even though it is more of an explanation for Kimble being on the run than a major plot point in most episodes. Gerard only appears in about a third of the episodes, the one-armed man in only 10 of 120 episodes.

The series also distinguished itself by having an ending. After four seasons on the run, Kimble finally catches up with the one-armed man with

Gerard there to witness it and Kimble ended up going free after a two-part series finale. The final episode was the highest rated episode of television at the time.

The Fugitive offered a very workable model for making an anthology series function as a continuing story through a simple overarching arc. A series like *Kolchak, The Night Stalker* (1974–1975) demonstrated the superiority of this form, when the straight horror anthology audience had dropped off after the heyday of *Twilight Zone, The Outer Limits* (1963–1965) and *Suspense* (1949–1954). Instead of just a host, Carl Kolchak provided audiences with a fully participating surrogate in the title character, a beaten down newspaper reporter who stumbles into weekly monsters or supernatural events. Viewers might prefer a homicidal monster to an episode with a ghost, but they'd still tune in to watch Kolchak's continual struggle against malign supernatural forces, on one hand, and his editors, the police, and political establishment forces that obstructed or censored him, on the other.

It's not surprising that Chris Carter, creator of *The X-Files* (1993–2002), continued to point to the inspiration of Kolchak and attempted to reboot the series with a younger, sexier cast. The revamped series, which did not find an audience, blended Kolchak with *The X-Files'* late 20th-century paranoia, but also moved the anthology further into a continually advancing, long-arc story.

The X-Files and Mythos in Longform Television

The Fugitive's use of a long arc to turn an anthology series into a longform television story comes to be a fixture of most dramatic series by the 21st century. Dramatic series tend to have a big mystery—generally a conspiracy—running along under freestanding episodes. This big mystery has come to be called a Mythos, from the use of the term by *The X-Files* writers and fans. The term actually comes from the "Cthulhu Mythos," a series of thematically related short stories by early 20th-century horror/sci-fi author H. P. Lovecraft and his followers.

In *The X-Files* context, the Mythos began as the long arc—Kimble's pursuit of the one-armed man while himself pursued by Girard in *The Fugitive*—the running story that disguised the show's identity as an anthology series, but it came to stand for the separating of the longform series into two parts: stand-alone episodes and Mythos episodes that advance the long-arc mystery.

Agent Fox Mulder (David Duchovny) works on the FBI's X-Files, the unexplained cases that he believes relate to the paranormal. Dr. Dana Scully (Gillian Anderson) is the skeptic assigned as his partner to keep him in check. The long narrative arc that spans the whole *X-Files* series is the gradual revelation of an incredibly complicated international government conspiracy with aliens. Mythos episodes are interspersed among the stand-alone paranormal or monster episodes that dominate the series. These episodes would have been familiar to Kolchak, as would the inevitable suppression of the evidence that Mulder finds.

The X-Files' Mythos becomes so complicated that shorter multi-episode arcs could appear in which pieces of the conspiracy were revealed, and specific mysteries could be solved without endangering the broader mystery. To further complicate the longform story, Mulder's mysteries are sometimes balanced by Scully's Mysterium. These stories of the supernatural world occur in the context of Dr. Scully's Catholicism, leading to the spiritual paranormal, especially a religious millennialism that included episodes in which Scully attempts to protect a girl who might be the direct descendent of Jesus of Nazareth. The increasingly End Times–based religious arcs spun off into their own series, appropriately titled *Millennium* (1996–1999) in which Frank Black (Lance Henrickson), a former FBI profiler, works with and against the Millennium Group to attempt to save the world from apocalypse going into the new century, while also solving—or at least addressing—supernatural single-episode stories.

The X-Files and *Millennium* are both built on the foundation that Roy Huggins laid with *The Fugitive*, but the anthology series episodes are increasingly subsumed by complicated and detailed sets of season-long and series-long arcs. For all the importance of global conspiracy, those long arcs are energized by soap opera emotional logic. *Millennium* gradually becomes the single long story of Frank's attempt to save his family while saving the world.

The government-alien conspiracy of *The X-Files* is grounded in the personal crises of the childhood abduction of Agent Fox Mulder's sister. Family secrets are inseparable from interplanetary conspiracies, as Mulder's father and then his mother are revealed to be complicit in both the alien abduction of his sister and the broader conspiracy. That conspiracy is embodied in the shady manipulator, the Smoking Man (William B. Davis). When it is revealed that the Smoking Man was a friend of Mulder's parents and might be Mulder's biological father, the personal and the political become inescapably collapsed. While not as central to the series, Dr. Scully's long arc relates through her family and her religious faith to the non-alien supernatural of the religious world. The show also contains the important but conventional series-long slow arc of romantic tension between the two lead characters.

The X-Files balanced its episodes and long arcs in a fundamentally different way than had been done before. Over its nine seasons, episodes come to be broken into Mythos and Non-Mythos stories by creators and fans; leech people and circus freaks and small cannibal towns, succubi, people who would not stay dead, homicidal incestuous families, and deformed Cher fans all appear in freestanding episodes, with no reference to the Mythos. The Mythos episodes, by contrast, tell a connected, advancing story that requires viewers to know what has happened in the past to understand who the characters are and what is going on.

Part of the series requires no background. Part of the series requires an ever-expanding body of knowledge. The division is so stark that, among the various DVD releases of the *The X-Files* at the height of its popularity, viewers could choose Mythos or Non-Mythos DVD sets, depending on whether the customer preferred to watch *The X-Files* as chapters of a serial novel or to just watch a collection of scary monster stories.

For its first three seasons, *Fringe* (2008–) followed *The X-Files*' model (and vocabulary), presenting a mix of freestanding and Mythos episodes but, after apparently being canceled for low ratings, and then un-canceled after Fox reviewed TiVo viewership and DVD sales, the creators of the show announced that they would reverse the traditional logic of network broadcasting and apply narrowcasting logic to the series. Since their targeted viewers were the devoted fans of TiVo and DVD sets, *Fringe* abandoned the stand-alone episodes and began their fourth season doing only Mythos episodes. To put it another way, they explicitly turned the episodes of a network television series into chapters of a single longform novel.

Chris Carter's *X-Files* and *Millennium* demonstrated two extreme ways to develop a long form story from an anthology series. The first is to divide a series into one set of freestanding episodes and one set that are chapters in a single longform story; the second is to simply let the longform story take over.

The anthology series wasn't the only model for a longform series, and it's worth taking one more look at how the soap opera could be applied to prime-time television.

Twin Peaks: Soapy Cop Show with Art

If people would tune into cable television to see movies, then why wouldn't they tune in to network television to see a television show created by a hot film director?

That was the reasoning behind ABC programming the David Lynch/ Mark Frost series *Twin Peaks* in 1990. In 1986 *Blue Velvet*, Lynch's blend of disturbing art house surrealism and Hardy Boys/Nancy Drew detection, somehow made a widespread connection with American culture. In the afterglow of the surprising mainstream success of his film, Lynch was offered the chance to create a show that reflected his personal combination of high art tragic and slapstick comic for network television.

As Stephen Bochco's *Murder One* (1995–1997) would do five years later, *Twin Peaks* followed one murder mystery for the whole first season but it also offered much trademark Lynch weirdness woven through the vaguely retro world of slow pacing and foreboding moody music of the classic soap opera, albeit with a self-aware twist, combined with mystical mood and surreal moments.

Including recurring visions of a backward-talking dancing dwarf.

The original surrealists took great pleasure in watching popular genre movies such as Louis Feuillade's 1915 *Les Vampires* serials, the *Fantomas* films (both series about master thieves moving in and out of the night windows of Paris), and genre films such as the original *King Kong* (1933), reveling in them as works made unintentionally surreal by the unconsciouses of their directors and the awareness of their audience. Lynch builds on this tradition by finding—and making obvious—the surrealism in the formula entertainments.

At the heart of *Twin Peaks*' unsettling assemblage of television conventions are two clichéd characters out of television cop story genre—the straight arrow overwound FBI agent, Agent Cooper, and the local underwound sheriff, Harry S. Truman, who also happen to be rough stand-ins for the show's two creators, Lynch and Frost. Together, these law enforcement officers navigate an explicitly soap opera world to solve a season-long arc investigating the murder of homecoming queen Laura Palmer. With film talent creating the series, and film directors brought in to direct individual episode, Lynch and Frost created a more cinematic series on network than cable could offer. That was enough for the initial audience, but ratings dropped after Laura Palmer's murderer was revealed. The introduction of a second season-long arc featuring a taunting serial killer, and David Lynch himself as Agent Cooper's comically deaf superior officer, did not hold on to the audience and led to cancellation.

Twin Peaks did significantly expand the vocabulary of tone on television, demonstrating how effectively mood and ambiguity—or cinematic and artistic thinking—could transfer to television, even commercially interrupted network television storytelling. If nothing else, the show gave viewers the spooky visual iconography of powerful high-beam flashlights in the woods at night that is now a visual cliché of suspense.

In a very different way, *Twin Peaks* was repeating what Bochco had done with *Hill Street Blues*, albeit simplifying Bochco's multitude of story arcs to the simpler and slower story arcs of classic soap opera. Lynch's goal was to combine a police investigation with a soap opera, while Frost wanted to create a Dickensian story that could continue forever. (While the show is always identified as David Lynch's creation, Lynch took off to make the film *Wild at Heart* [1990] after the second episode, and Mark Frost was guiding writer for most of the series.)

Where Bochco stripped out the essence of the soap and grafted it into the cop show structure, *Twin Peaks* doubly foregrounded the soap opera as the central element of the series. First, the explicit storylines were reinforced by daytime drama visuals and music. Lynch and Frost had studied the early prime-time soap, *Peyton Place* (the 1957 adaptation of the record-breaking best-selling 1956 novel), in preparation for writing the series. The original intention was that the mystery of the murder of Laura Palmer would be the hook that attracted audiences and that the mystery arc could give way to the soap opera once the audience knew the show's characters.

Irony and Postmodernism in Longform Television

Twin Peaks also highlighted its identity as a soap opera by having characters regularly watching *Invitation to Love*, a fictional soap opera within the series. Besides being one of the first times that American television characters were regularly shown watching American television, the soap opera within the soap opera commented on the series by foreshadowing or contextualizing events. The arrival of Laura Palmer's look-alike cousin (played by the same actress, Sheryl Lee) is preceded by twins with diametrically opposite personalities on *Invitation to Love*. Other events on the soap seem to be attempting to warn *Twin Peaks* characters on how to react to the plotlines that entangled them.

In the critical vocabulary of the day, the show's self-aware artificiality is a tactic of postmodern self-referentiality that creates an ironic distance for viewers; television formulae are deconstructed through the creation of a simulacrum of the banal genre conventions, leading informed viewers to watch the series not as a genre show but a meta-genre show (that is, a comment on the genre while taking its form), turning a television show into a work of avant-garde art. For others, *Twin Peaks* was just another slice of the David Lynch weirdness that made a murder mystery inside a soap opera a strange but somehow entertaining experience.

What *Twin Peaks* did with conventions of genre television compares to what Dennis Potter did with detective, sitcom, and musical genres in *The Singing Detective*; Potter developed his approach from modernist literary traditions, Lynch built on the same traditions from the world of visual art, especially surrealism, but his work was experienced as being postmodern. The key distinction in an unnecessary attempt to categorize the show may be whether the show is meant to be taken as sincere or as ironic.

Lynch seems quite capable of holding both attitudes toward his material, simultaneously. There is an absurdity to the soap opera elements but they are treated very seriously, creating a pleasing dissonance that is characteristic of Lynch's work. Dennis Potter's *The Singing Detective* treated its genre elements in a similar general way, although to different effect, and there is clearly an aesthetic kinship between Lynch and Potter. At the height of Potter's acclaim, there was talk of Lynch directing a Potter script.

Twin Peaks' dream sequences/hallucinations/visions are nothing like those of *The Singing Detective*, in part because they originate from a source outside the dreamer, a sense of dream that is Jungian in its offer of mythic guidance rather than Potter's Freudian dream structure based in childhood memory and sexual shame. Similarly, most of *Twin Peaks* has a soft-focused, packed-in-cotton tone to it that heightens the unsettling biological blood-clot clarity of the show's alternate reality. Potter explored the bleed between alternate consciousnesses where Lynch saw characters move in alternate dimensions.

It is easy to forget that the first season of *Twin Peaks* was only eight episodes, roughly the length of a British television series or some cable series, and the show provided closure at season's ending while planting new themes for the second season. The series was a filmmaker and film writer telling a murder mystery on television, using the associative visual and aural language Lynch had developed for art films on the most mass of American media. The dissonance was disturbing and, surprisingly, temporarily successful. A telling footnote to the series is that Lynch recut the *Twin Peaks* pilot as a feature film, with the mystery solved, for direct-to-video release outside the United States, an act that suggests how little artistic distinction he made between storytelling through theatrical film, television, and video.

In 1999, ABC again considered making a television series with Lynch. Going into the new century, American audiences had shown they were ready for more challenging longform television, but ABC decided they probably weren't ready for the series called *Mulholland Drive* on network television. Lynch shot some extra footage and turned the pilot into the feature film—just as David Chase thought he would have to do with *The Sopranos* pilot. (*Mulholland Drive* became one of Lynch's most successful films and Chase's series got picked up by HBO and became very successful in its own right.)

ABC probably made the right business decision regarding *Mulholland Drive*. The two-hour (with commercials) pilot exhibited what becomes an increasingly fragmented sort of storytelling, with scenes of characters out of context that set mood and implied eventual connection and explanation, with possible juggling of time sequence and Jungian sorts of dream symbolism, but without the guide ropes of a familiar genre narrative that *Twin Peaks* offered. Lynch intended to create the series with multi-episode arcs as long thematic chapters, but Lynch's storytelling in the new century left clear—and network—narrative behind.

Lynch seems to have implicitly endorsed that judgment in making *Inland Empire* (2006) as a television show that could never air on network television. The story is deliberately unclear but it plays out around Laura Dern who might be an actress in Eastern Europe becoming the character she is playing in a film, or might be dreaming she is an actress, or might be a character in history's longest-running radio drama, *Axxon N.* Time sequence, levels of reality, and geography shift in different directions without notice. At one point, Dern's character may have a doppelgänger, may be seeing herself from an alternate time, or may be seeing the character she is playing within the film. At times, the film is interrupted by a television series about bunny-headed people.

The closest model for the storytelling in the film is the grammar of dream that was already dominating *Mulholland Drive*, with identity and causality flowing and pooling in different patterns over the film's three-hour length. The film was Lynch's first shot at digital media. While it received a theatrical release, Lynch saw *Inland Empire* as something to be watched on television from a DVD, having mixed the visuals and audio for the television, not the movie screen. Those who prefer to think metaphorically might consider the parallels between the opening of the film, in which we hear the beginning of a radio show that we see is being played from a recorded disc, and the fact that the audience is simply updating the same media relationship as they are watching television/film from a DVD disk.

The three-hour DVD represents a new sort of television, coming from the film end of the spectrum; a longform story configured for DVD-television viewing. Asked about the length, Lynch explained that he made *Inland Empire* with the idea that viewers would watch for a while, then stop and go away, to come back later and watch more, the way someone would read a book.

While Lynch lives in a world of visual rather than textual metaphors, *Inland Empire* is a DVD novel, approaching that form from the cinematic side of the bed.

Effects of Longform Storytelling on Villains

In single episodes, the hero needs only to locate and defeat a problem. In a longform story, a protagonist needs a worthy antagonist. Dramatic conventions push us toward that need for more rounded characters, hero and villain.

One of the side effects of the longform story is that the nature of television villains changes. In the first, episodic, season of *The X-Files*, the Smoking Man said four words and was seen but not heard on the periphery of events, but to sustain nine season, he had to be fleshed out. Over the years he received a family, personal crises and losses in love, a real and false history, a set of motivations and bosses to whom he had to answer.

The series that embraced the problem of creating villains equal to longform television storytelling was *Wiseguy* (1987–1990).

Wiseguy's response to the threat of cable movies was to create a series that built on *Hill Street Blues*' realism and soapishness but acted more like a feature film. The additional twist was that show's creators recognized that longform television offered them a length and depth of storytelling that a single film could not.

To enhance its difference from traditional television, the series was shot in a more cinematic style than most television of the time and made extensive use of young film actors and celebrities. Ray Sharkey, Kevin Spacey, Annette Benning, Ron Silver, Tim Curry, and Patti D'Arbanville, and musicians such as Billy Vera, Glenn Frey, and Deborah Harry gave the show an edge not associated with television. Jerry Lewis and Paul Winfield used arcs on the series to show different sides of their public images.

The series took the popular film drama of tormented undercover police officers such as *Serpico* (1973) and *Prince of the City* (1981) and moved them onto television. The television cop show became a place to explore issues of morality and compromise in a way that it never had before. *Wiseguy* revealed—and reveled in—the emotional and psychological damage that resulted from undercover agent Vinnie Terranova (Ken Wahl) living a double life of lies and betrayal, committing crimes to punish criminals.

Like the disguised anthology model that lay at the heart of *The X-Files* and *Millennium*, *Wiseguy* features an undercover Organized Crime Bureau (OCB) agent, who is dropped into a series of different stories with new criminals and then moves on. The adjustment that the show made was that Terranova dropped in for long multi-episode arcs running up to 12 television hours. The longform stories of these arcs inevitably found Vinnie Terranova developing the attachments and complications that the drifter anthologies avoided.

The first season consisted of two parts—a 10- and a 12-episode arc. Viewers saw Vinnie struggling between competing loyalties and moralities as he pretended to be a criminal to do the job of betraying people he had more in common with than he did with the people who signed his checks. The show's discrete multi-episode arcs provided a new business model for longform television by offering viewers a second chance at midseason to enter the series from the very start of a new story. CBS was not persuaded, though, and continued to press for the show to be closer to a typical collection of stand-alone episodes. The second season shortened the multi-episode story arcs to 4 episodes (White Supremacists), 5 episodes (Garment Industry), and 7 episodes (Music Industry) now separated by stand-alone episodes, essentially settling into what *The X-Files* would later do. The third season presented three short arcs, and the final season only one.

Wiseguy: The Longform Sympathetic Villain

The first arc was 10 episodes long and developed the essential structure of *Wiseguy*. Vinnie joined Sonny Steelgrave's (Ray Sharkey) Atlantic City operation, doing his job while also looking for revenge for Sonny's brother having killed Vinnie's mentor. Terranova soon experiences conflicting emotions as he protects Sonny, as part of his job but also out of impulse, because he likes the guy. His simple job continues to get complicated by the emotions and moral ambiguity that were left out of traditional television cop shows. Over the 10-episode arc, Vinnie develops increasing doubts about his work as he becomes close friends with Steelgrave. The relationship develops because the two men share the same strong sense of honor which allows them to do things that most people would consider immoral. Unfortunately, this means Vinnie must betray Sonny.

Wiseguy's themes are not single episode themes. The 10 episodes gave the writers room to consider different aspects of conflicting moralities. Longform television is often compared to film, but the Steelgrave story arc was roughly the length of five movies. That room allows the creation of a genuine dramatic arc. While Vinnie's arc is circumscribed by his continuing role in the series, the revelation of Sonny, the sympathetic villain, is a gradual, multilayered transformation of character and relationships that could not be done in the time frame of a film and probably not onstage. Only the novel can tell a story of comparable length. More than *Twin Peaks*, the less arty *Wiseguy* showed that multi-episode arcs on television had the potential to surpass the storytelling possibilities of a single film.

The final episode of the Steelgrave arc gave the audience the law enforcement raid on the mob meeting they'd expect from television convention, but Steelgrave and Vinnie end up locked into a movie theater that had closed down for the season. While not as obvious as *Twin Peaks*' *Invitation to Love*, an appropriately self-referential setting for considering the inadequacy of the standard Hollywood depiction of cops and mobsters. The two men alone in the movie theater fills most of the episode.

Sonny knows Vinnie has betrayed him, that the police have shut down his operations and he'll spend the rest of his life in jail. The two men beat on each other until they collapse in exhaustion. At which point they begin to talk and work together to try to escape the movie theater, then give up and start drinking alcohol conveniently left behind. Steelgrave goes to a jukebox and plays "Good Lovin'" by the Loving Spoonful. The entire song plays as Vinnie watches Sonny trash the concession stand, pointedly singing along that "all I need is love" to the man he thought of as a brother. The next song on the jukebox is Moody Blues' "Nights in White Satin." The entire song plays as the two exhausted men lie bloodied in white shirts, each holding a bottle of liquor. Sonny interrupts to ask Vinnie how he's going to remember him but, mainly, the song plays as the camera slowly cuts back and forth between increasingly closer shots of the two men. Lyrics like "just what the truth is, I can't say anymore" and others about being alone, being unable to defend one's actions, the inability of others to understand dominate the audio track before the song builds to its repeated claims of love.

For two minutes, no one speaks while the song finishes and we watch the men think their separated thoughts.

Two minutes was an incredibly long time for a television series of the time.

This is an example of cinematic—not television—storytelling in its day. The men avoid directly addressing the only thing that matters between them, and the audience watches and listens and decides for what each man is thinking as the song fill the gaps as they come to terms with the betrayal of a loving relationship (*phileo* or *storge*, not *eros*, for those concerned about such things). The emotional bonds between men were acknowledged but not examined in traditional television, definitely not to this level. As we watch their angry and guilty confusion as their roles revert from close friends to cop and criminal, themes planted in the previous nine episodes are being paid off in a demonstration of just how messy the world of this story is.

The sequence is also a bit audacious because the song is too "on the money" to put on the soundtrack of the scene. The song addresses the inability to know what is true, inability of others to understand your feelings, people trying to share thoughts that don't hold up, how you are responsible for who you are, and passionate love. Except the song is not on the soundtrack, Sonny played it on the jukebox for Vinnie to hear. It is Steelgrave, not *Wiseguy*, that is too heavy-handed in selecting music.

(Before you rush out to get the DVD set of the first season of *Wiseguy*, be warned that, like *Northern Exposure, WKRP in Cincinnati*, and some of the *Miami Vice* DVD sets, the producers did not clear the music rights for the DVD sets, so the specific songs that were so central to those series has been replaced with filler music. You can probably still see and hear the scene in its original form by searching "Wiseguy—Nights in White Satin" on YouTube. Comparing the generic music on the DVD set with the original use of the song clearly shows what using a recognizable song adds to a television episode.)

When the police do arrive, Steelgrave chooses an end that calls to mind James Cagney in *White Heat* (1949) and Vinnie ends up in a sanitarium. Television writing of *Wiseguy*'s day continued to follow the rule of no change. Vinnie stays roughly the same throughout the series, a man tormented by his inability to live with the betrayals and deceit that are also his life but, within each multi-episode arc, he goes through a "non-television" dramatic arc of transformation that would be taken for granted in a film or a novel, only to reset for the next arc.

Wiseguy: The Longform Charismatic Villain

The second, 12-episode arc is what *Wiseguy* is best known for; Terranova infiltrates and betrays the operations of Mel Profitt (Ken Spacey), a mentally disturbed billionaire crime lord, drug dealer, arms dealer, addict, incestuously involved with his unstable sister, Susan (Joan Severance), who becomes Terranova's lover. Spacey's performance seems too big for television of the time but it ends up establishing the skewed reality of a complex story in which each of the main characters has multiple dimensions that gradually expand through the three months—or 12 hours/chapters, and the eternal verities of network television turn out to be neither. Profitt doesn't make it to the last episodes of the arc. By that point, he is just one of many who are dead, but he is killed at his own request, by his sister, after which she goes insane.

In the 11th episode, the story returns to secret locations and private armies planted in the first episodes, but neglected since the beginning of the arc. The Profitts are out of the picture but Lococco, their personal hitman who was also an undercover CIA agent, takes over Profitt's operations. Lococco has been using Profitt's money to fund a secret army for the CIA to install a puppet dictator in control of a Caribbean island. The final episode of the arc has Lococco testifying in televised congressional hearings and his corrupt superiors lying their way out of any responsibility. Vinnie, who had begun the arc in group therapy, tries again to quit the OCB but is given a vacation instead. The simple crime boss story continues to open up to reveal new depths of corruption and complicated moralities, as Profitt's drug trade supported his international arms dealings, which involved him with illegal CIA plots. Unexpectedly for network television in 1987, U.S. government bureaucrats in the CIA, FBI, OCB all are motivated by self-interest, ego, and outdated anti-communism.

This compressed summary probably sounds fairly ridiculous but the revelations are made gradually over 12 episodes, as layers of a simple drug operation are repeatedly peeled back to reveal even more corruption and damaged people than were previously understood. What makes it believable is that the characters are written and performed in a disarmingly exaggerated style. Spacey's performance as Profitt deepens through layers of charisma and humor, brilliance, insanity, pathetic dependence and vulnerability, and pure tactical amorality. Reinforcing the arc's plot and mood, the contradictory layers of Profitt's personality are steadily peeled back to present an increasingly complicated vision of a mentally unbalanced criminal business genius. This is a more challenging vision of a villain than television had previously had time to show viewers. In a striking statement for American television of the time, Profit at one point challenges his critics by declaring his principles: "You think poor peasants care who's ramrodding them? The glorious flag of democracy or the tyranny of communism? They starve either way! I don't want control of your politics! I'll make money off your chaos and your madness . . . but I DO. NOT. TAKE. SIDES!"[2]

By the time of his death, Profitt has revealed himself to be the most honest and clear-sighted character in the show, while also being insane and ruthless. With Sonny Steelgrave, *Wiseguy* forced the audience to empathize with Vinnie's respect for the villain, if not the villain himself. With Mel Profitt, the villain is, in many ways, the most admirable player in a world of corruption that only he really understands. This is not what viewers expected on network television in the same season that *The Cosby Show* (1984–1992), *Family Ties* (1982–1989), *Cheers* (1982–1993), *Murder, She Wrote* (1984–1996), and *The Golden Girls* (1985–1992) were the highest-rated shows on television.

Attractive Villains and Unattractive Heroes on Longform Television

With longform television the villains begin to change. And so do the heroes.

In an episode-based series like *Baretta* or *Columbo* (1971–2003), to use an adequate and a great example, we learn very little about the hero. Audiences are given factoid knowledge, he keeps a parrot, he drives this car, he is smarter than he appears, he tells strange stories about his wife whom nobody ever sees. Mainly we see the character do his job. It is a testament to the power of episodic television writing that so many people could care so deeply about Columbo, a character about whom viewers know so little.

Earlier I mentioned the detailed backstory information about *The X-Files'* Smoking Man that fills dozens of pages at online sites. After 69 episodes, each one-and-a-half or two television hours in length, what we know about the man, Columbo, fills about a page. It is mainly a listing of the brand of cigars he smokes, where he keeps his gun, the fact that he gets seasick and brings hard-boiled eggs to morning cases, and has a shoe size of 10-1/2 or 11. After 32 years on the air, it remains unclear if Columbo has children.

More is written about how Peter Falk got the role than about Columbo's life.

Columbo is generally considered one of the greatest characters in television history and we know almost nothing about him. Traditional television writing emphasizes the fact the "action is character" and leads viewers to empathize with the hero's values, his intelligence, and his wit as we see him demonstrate those attributes in solving his cases. But his broader life goes unexamined.

Longform television, with its soap operatic roots, leads us to learn much more about characters. As is often noted, a straight-arrow character is hard to write because he is boring. Moral complexity makes for more interesting characters. In the same way, moral ambiguity is a side effect of longform television. And literature. Network television gets attractive villains and ambiguous heroes—or antiheroes—when the stories extend beyond individual episodes.

With the Fox series *Profit* (1996), the pendulum swung too far too fast, but the series left us with the template for later protagonists like Dexter Morgan, Al Swearengen of *Deadwood* (2004–2006), Vic Mackey of *The Shield* (2002–2008), and possibly Don Draper from *Mad Men* (2007–).

The name of the eponymous protagonist of the largely forgotten series, *Profit* (1966), was intended as an homage to *Wiseguy*'s Mel Profitt. The show puts the attractive sociopath willing to do anything to get what he wants at the heart of a dark-hearted modern soap opera world.

In the early years of Fox Television, the network attempted to build an audience by targeting audiences that were being underserved by the existing Big Three. Adding "edge" to their programming was seen as a way to attract younger and urban audiences.

Profit was an example of edge.

While *Profit*'s sociopathic hero might, technically, be seen as one with *Dallas*'s JR Ewing, a soap opera villain the audience enjoys cheering for, Profit does things JR never did, including a lot of murders and (possibly) sleeping with his own mother. Mainly Profit attempts to kill and manipulate his way to the top of the G&G Corporation.

Profit's story actually foreshadows Don Draper's on *Mad Men.* An abused child grows up and reinvents himself by taking another man's identity. He uses it to get to the heart of the corporate world, where his natural abilities allow him to thrive. The difference is that Profit is a homicidal sociopath. Draper is an ad man.

Having been raised in a cardboard G&G box, watching G&G commercials on the television that was his only caregiver, Profit's goal is to make his way to the top of the G&G Corporation by any means necessary. Each episode of *Profit* features a story of an immoral intrigue that ends with Profit one step closer to the top of the corporation.

Wiseguy constantly raised questions about the limits of television morality, offering a range of conditional and situational moralities that do not satisfy the protagonist or antagonists. Profit commits immoral acts, but they are committed in an immoral corporate environment where profit and power are the only absolutes, values Profit is willing to share.

Profit was canceled with episodes left unaired. Two reasons have been suggested; complaints from viewers about the lack of punishment for a lead character thriving through immoral behavior, and complaints from corporations about the suggestion that corporations commit or encourage unethical behavior. While the highly popular JR Ewing destroyed many more innocent people than Profit, he was good television because he was motivated by familial love and jealousy. Profit did bad things without rationalization.

In retrospect, the series has been pointed to by many as the show that broke from the idea that television has to teach moral lessons to its audience, paving the way for successful cable series of conflicted morality like *Dexter* (2006–), *The Sopranos* (1999–2007), and *The Shield*, where the audience comfortably watches lead characters kill and commit horrible crimes on their way to success.

The final evidence of the triumph of the appealing villain as hero on network television may be found in NBC committing to a 13-episode series about Hannibal Lecter for the 2012 season, the cannibalistic serial killer from Thomas Harris's novels.

The first time that viewers had the chance to see the full season of *Profit* came when the complete series was released on DVD in 2005, making it an example of another step toward the DVD novel. Too late, the final episodes on DVD made the series' black humor more obvious, guiding viewers toward a subtly different understanding of the story than the partial series they saw on television.

Chapter Five

Premium Blend Genre and Longform Network Television

In Which We Take a First Look at Dramedy and Consider *Gilmore Girls*, *House*, and *Veronica Mars* as Models of Successful Longform Network Television

Genre is a bookstore problem, not a literary problem.

—Rick Moody[1]

Blending two established television genres to create something new has been a tactic since the start of longform television. *Hill Street Blues* (1981–1987) set the model for blending the soap opera with something else, which quickly became the norm for network television. Adding comedy to drama, like soaping up a cop show, in theory doubles the appeal to audiences—it's not just a cop show, it's also melodrama; it's not just a detective show, it's also a romantic comedy. And once the lawyer-soap is established, a lawyer-soap with comedy can take it even further.

Some of the most critically successful longform network shows of the turn of the century offered a blend beyond the soap-plus-one approach. *Gilmore Girls* (2000–2007) and *Judging Amy* (1995–2005) combined the mother-daughter drama with another genre, screwball comedy and the law show, respectively. *Veronica Mars* (2004–2007) combined the high school soap with film noir. *House* (2004–) combined the medical drama with the classical detective mystery and a bit of the acerbic British sitcom.

In the realm of blending comedy with another genre, there are a surprising number of subtle variations. The descendents of *L.A. Law* (1986–1994), including *Ally McBeal* (1997–2002), *Boston Legal* (2004–2008), and *Harry's Law* (2011–), the lawyer-soap-with-laughs are probably the most prevalent, but there are the lighter variations like *Suits* (2011–) and *Franklin & Bash* (2010–); there's also the detective show with laughs—*Moonlighting* (1985–1989), the witness protection show with laughs—*In Plain Sight* (2008–2012), the ex-CIA agent caper series with laughs—*Burn Notice* (2007–), the TV show with laughs—*Sports Night* (1998–2000),

Studio 60 on the Sunset Strip (2006). These blendings of genre with humor end up forming one of the new conventions of American television writing; the non-comedy provides the long arc, the comedy provides the episodic closure.

Comedy lives in the moment and in immediate reactions so, besides the occasional explicit joke, the humor in these shows tends to be a narrative pattern of reversal that becomes the feel-good closure of the episode, reinforcing the closure of a more serious case. A bad guy threatens the viewer's surrogate and, at the end of the episode, is humiliated. One of the good guys reveals the other good guy had too high an opinion of himself, and healing laughter closes out the episode.

This is a deliberately oversimplified explanation, because the lack of humor in an episode may also be the closure in a more serious story, so sometimes a show ends with an unfunny realization by the good guy, or the bad guy temporarily triumphant. But, generally, the rhythm of the individual episode on American network television is toward entertainment and leaving the viewer feeling good at the end, whether it's fiction or the nightly news. In other cases, light humor acts as a reassuring guide through the serious story. A viewer watching *White Collar* or *Suits* or *Franklin & Bash*, for example, can enjoy the light comic attitude of the storytelling that promises a happy ending, even as they also become involved in the supposed dramatic tension of whether the good guys will win the case.

Those are the single-episode gratifications.

But the shows also have a soap opera–derived long story that becomes more dominant in the viewing experience when watched in sequence or in a binge viewing. Will the federal agent heal her family and finally find her father? Will the former con man locate the music box that leads to the Nazi treasure and defeat the evil conman nemesis who dogs his footsteps and threatens his friends?

Long arcs are dramatic, not comic. This is not to say that any television creators actually sit down and formally think through blending different genres, although network executives might encourage it as a marketing strategy. Blending genres, and blending comedy with drama, is the way things are done now. So not blending in humor is the decision more likely to require conscious decisions. Creative people try to make something that "works" and formal structural analysis is rarely part of the process.

Similarly, once *Hill Street Blues* had established the soap opera injection into prime-time storytelling, it simply became the new normal. Creators did not discuss needing to add soap opera elements; instead they worried about character backstory and arc, the personal storyline to balance the plot.

Gilmore Girls

Gilmore Girls is a series about a single mother, Lorelai Gilmore, who had a daughter, Rory, when she was an unmarried high school student of 16. The series begins with the daughter, now 16 herself, and the mother having turned her back on her wealthy parents building a life for herself and her daughter on her own.

The series is very much an auteur vision with the voice of creator Amy Sherman-Palladino. The pilot episode is one of the most impressively comprehensive, yet well-written pilots created. Without the use of what Terry Gilliam calls a "Mr. Exposition"—someone who seems to be having a conversation but lays out everything the audience needs to know—the pilot outlines the entire series; the backstory, the attitude of the show, the secondary arcs of the show, as well as the major transformations within the long arc that will have to have happened by the end of the show's run.

The series starts with the daughter, Rory, transferring to an expensive private high school, which requires the mother, Lorelai, to borrow money from her estranged wealthy parents in exchange for weekly family dinners. The series will follow Rory through high school, then through college, and end as she sets off into the world. The series will also follow the gradual restoration of the estranged family into a loving group. The series will follow Lorelai as she becomes financially independent and successful and therefore able to approach her parents as equals. And the series will eventually put Lorelai together with the obvious love interest of Luke, the working-class diner owner. (Soap opera fans would have picked up on the inevitable romance immediately, since the couple's names are Luke and Lorelai and the high point of daytime drama ratings was the 1981 marriage of Luke and Laura, the young lovers on *General Hospital*.)

The series is worthy of study for the way it balances the series-long story arcs, the season-long stories, and the episodes. Partly this was accomplished by blending two genres, primarily the prime-time mother-daughter soap opera like *Providence* or *Judging Amy*. The former added a medical show, the latter a law show.

Gilmore Girls blended that mother-daughter soap with the screwball comedy films of the Depression era (*Bringing Up Baby, The Awful Truth, His Girl Friday*, etc.). The screwball fast-talking witty banter and inside jokes, and a small town filled with screwy supporting characters provided the natural material for freestanding episodes, while the melodrama and romance could be developed in the long arcs.

Sherman-Palladino's reconstruction of screwball in a mother-daughter melodrama is an impressively complete project. Each episode has a melodramatic storyline built around family, dating, or career. But each episode also has a wacky plotline involving one of the many screwy characters in the town of Star's Hollow. The serious storyline is part of a multi-episode arc that will not be resolved in a single episode, but the wacky story will resolve itself by the end of the episode leaving viewers smiling, even if the continuing story did not. Where the witty banter of the classic screwball comedies was a sign that the romantic couple belonged together, the shared banter of mother and daughter sets them apart from others and bonds them together as the series' primary couple; Luke and Lorelai's playful but thinner banter marks their romance as the second most important relationship in the series.

The show that may have most successfully built on the *Gilmore Girls'* model of adding screwball is *Bones* (2005–), which runs a muted screwball variation on a *CSI*-style show where the scientific team investigating murders is made up of quirky characters who can provide comedy for the freestanding episode but are still rationalized enough that they serve the soapy concerns of the longform stories (hurdles to romance, hurdles to marriage, hurdles to family happiness, inappropriate patterning on a father figure who is also a fugitive murderer) so that the continuing story arcs are often carried by the supporting characters rather than the will-they-or-won't-they screwball pairing of the impulsive FBI agent and the so-rational-she's-screwy forensic scientist in the lead. The key distinction from *Gilmore Girls* is that the comic elements of the characters are generally in support of the murder mysteries and dismembered corpses.

Gilmore Girls is also worthy of study because, as the CW replaced the WB network that aired the series, the studio essentially took the show away from Sherman-Palladino for its last season and turned it over to a former show runner for *Spin City* (1996–2002) with the intentions of saving money and increasing ratings.

Under Sherman-Palladino, the show had developed a very distinctive set of comic voices that distinguished characters and allowed for different sorts of humor based in the characters. Most obviously, this included the inside jokes and topical references in Rory and Lorelai's banter but also in the stodgier wit of Lorelai's father and the cutting humor that leaked through in some of her mother's comments. Luke's voice was based in his practicality in contrast to Lorelai's fantasies. Taylor's humor was based in him never understanding there was a joke in the vicinity.

With the new producer, the show did not become horrible, but the voices blurred. Suddenly, everyone had at least a bit of a previously unshared Gilmore Girl–ish wit.

The voices of the Gilmores were also diluted as the new CW started to monitor online feedback to adjust the scripts. The goal was to encourage a *Gilmore Girls* community but also to find a way to give them what they wanted. Not surprisingly, the targeted young women commenting online wanted the characters to act more like "real" characters do—which generally meant like characters on other shows, wanted more romance, and didn't want to hear jokes they didn't understand that made them feel dumb.

So the final season was written to accommodate their audience, in the process replacing quirkiness with the predictability their respondents requested. Jokes were also made more accessible. Where *Pippi Longstockings* was the date movie in earlier series, the Audrey Hepburn film *Funny Face* is the date film in the last season. What might have been a reference to Gabriel Garcia Marquez's *Love in the Time of Cholera* in the previous season would be a reference to a Julia Roberts movie in the last season, the audience might not catch a Charlie Watts reference, so they'd refer to the more recognizable Mick Jagger.

Students of auteur television should watch the entire run of *Gilmore Girls* and see the way that a change in management shifts the essence of a show and the way that subtle shifts of tone change. The new producers did provide viewers with a series conclusion, although it left many fans dissatisfied, especially since Sherman-Palladino was not allowed to write the final episode, having been alluding to her planned ending for the series, and specifically her certainty about the final four words, since the series began.

To this day, she has not revealed the four words. (I suspect one of the four was "coffee.")

George R.R. Martin, a television writer and the author of the source novels for HBO's *Game of Thrones* (2011–) series, delivered a more sweeping condemnation of the last season of *Gilmore Girls* from the perspective of a writer: "The new person brought in to mop up during that last year just didn't have a good feel for the characters or their world, and after spending six seasons with the people of Star's Hollow, the final season was a trainwreck on a character level. A fairly decent series finale wasn't enough to remove the bad taste that the rest of that unfortunate season had left. Very little of what transpired was true to the vision I'd watched for the first six years of the show."[2]

Martin's complaint is essentially that of a longform television fan, expecting the voice of a show (or its creator) to remain consistent but also, after a seven-year investment, expecting the show to conclude in a way that brings the themes of the series together in a satisfying way. His comments were part of a larger defense—or explanation—of his thinking after his criticism of the last episode of *Lost* (2004–2010) caused an online kerfuffle. Martin went on to point out that what we expect from television series has changed. "Much of this has changed in recent years, of course. We're now getting more shows that feature an actual continuity, with episodes that need to be shown in a certain specific order to make sense. Shows that feature some central mystery that begs for solution, or some ongoing conflict that cries out for resolution. Such shows require an ending, whereas the older episodic model did not."[3]

Veronica Mars: Hard-Boiled High School

Veronica Mars failed to satisfy the network and attract an audience of the size they wanted, so the show was cancelled after three seasons. The short life of *Veronica Mars* is exemplary, though. Not just because it demonstrated the power of endings in longform television, but also because its success wasn't enough to allow the show to be different.

The show's roots were in a teen boy detective novel outline that Rob Thomas had sold to Simon & Schuster publishing. Thomas, author of several teen novels, writer for many television shows, and creator of the much mourned disguised anthology series *Cupid* (the quirky one from 1998 starring Jeremy Piven, as either a madman or the god of love, and Paula Marshall, as the psychiatrist and romantic advice columnist, put in charge of him, not the flaccid and formulaic 2009 reboot with Bobby Cannavale and Sarah Paulson. The difference between the two versions might be distilled to the fact that David Johansen of the proto-punk band, New York Dolls, did not appear as either the Greek god Zeus or a homeless man in the reboot.) *Cupid* created a structure for 100 freestanding love-story episodes and a longform screwball comedy love story between the Piven and Marshall, where the gender roles were reversed to present the woman as overly rational and controlling, the man as the screwy character in touch with love.

Veronica Mars pushed the long arc even further toward being a novel on television. Not surprising since Thomas created the story as a young adult novel, then wrote a sample script that was aimed at getting a premium cable job but ended up being picked up by the UPN network where the show became one of the network's highest rated shows and garnered much critical praise.

Seasons 1 and 2 told the story of Veronica investigating a single mystery established in the first episode of the season and solved in the last episode of the season. But each episode included one smaller mystery that was solved by the end of the episode.

The show achieved the most complete balance between serving the viewer who wanted to watch a single episode as a short story and the viewer who wanted to watch it as a chapter of a continuing mystery.

Veronica Mars combined two different successful genres, the high school melodrama and the family-centered hard-boiled detective story refined by Ross Macdonald from the Raymond Chandler tradition. Here the world-weary beaten-down gumshoe is a teenaged girl with ironic voice-overs detailing disappearances, rape, murder, addiction, and domestic abuse, while also slipping into the problems of high school cliques, bad dates, and "cool kid" parties.

Veronica had been part of the wealthy social circle before the show begins because her father was politically connected as the town sheriff but loses his job after he is blamed for failing to solve the murder of one of Veronica's wealthy friends, and Veronica is ostracized. The girl detective makes a new set of working-class friends, before gradually reentering the society of the wealthy with a new cynicism.

Where a later show like *Gossip Girl* centers on the conspicuously consumed lives of privileged teens in Manhattan, *Veronica Mars*'s landscape was the town of Neptune, California, divided between the fabulously wealthy and the working class. *Gilmore Girls* included issues of class, but mainly as signifiers for something else: Lorelai's parents were wealthy and judgmental, everyone else was working class and supportive. *Veronica Mars* repeatedly has to investigate issues of entitlement, deference, and resentment arising from class.

Veronica is raised by her father, who works as a private detective with Veronica's help. She also does her own investigations. Most prominently, she wants to solve her friend's murder and prove her father did nothing wrong, but that case also involves her finding out the details of her own drugged rape, which she is not sure she wants to know more about. Both cases throw her into the company of the wealthy, but as her suspects rather than friends.

The first season consists of Veronica's season-long investigation of the murder, while also navigating class differences with her former wealthy friends and new working-class friends. To use the vocabulary of *The X-Files* (1993–2002), every episode is a Mythos episode, because the Mythos of *Veronica Mars* is essentially conspiracy as soap opera. She must discover the family secrets behind the great crimes in Neptune and the forces that work to keep them concealed. But every episode is also freestanding, with Veronica helping her father on a case or freelancing a high school case on her own, tapping into this same aquifer of corruption in Neptune, but in smaller doses.

Viewers of season 1 had 22 chapters to learn about Veronica and the case by observing her in relation to her father, her old friends, her new ones, and to learn about events from before the series began. Individual episodes added to a sense of how she worked and what she cared about, as well as a broader sense of Neptune, a town of privileged wealth and working people and poverty, of multiple variations of racial relationships, of political loyalties and resentment. All the single episode details add to viewer's understanding the murder she was investigating, and the ways and reasons that people hid the truth, intentionally and unintentionally concealing the identity of the killer.

Central to the process of solving the case, Veronica has to gradually work her way through her resentment against friends who had abandoned her, to understand and empathize with people who were not like her, to learn what people do for love, for security and for family, even against their own best interests. One of the controversial elements of the first season storyline is that Veronica was drugged and raped and, in solving the murder, she must also discover what actually happened to her. Veronica eventually discovers that her rapist did not know that she was drugged and was told she wanted him to have sex with her, so she was essentially raped by proxy. But even with that excuse, it is clear that her rapist knew he was wrong in doing what he did, just not in the way she had thought. He isn't a villain, he's a manipulated person but still one who put his interests over hers. Finding the answer at the end of many false turns, there's no satisfaction, only knowledge.

This is not the sort of story that can be told in a single television hour.

Before the third season, UPN was sold to the WB whose executives were convinced the season-long story arcs kept viewers away, even if it was woven through the freestanding episodes. So the third season was plotted for three multiple-episode arcs, and almost all of the themes involving class disappeared from the show. After the first two arcs of eight and seven episodes were completed, the network put *Veronica Mars* on hiatus and replaced it with *The Pussycat Dolls Present: The Search for the Next Doll* (2007), a "reality" show competition in which young women did things such as posing provocatively in lingerie to win a position in the burlesque rock group.

The replacement for *Veronica Mars*'s strong young woman defined by her mind and her ability to act outside of stereotypes, attractive because she was unconcerned about being an object of sexual desire, could not have sent a clearer message of the direction the WB would be going. The remaining episodes of *Veronica Mars* were unconnected, freestanding episodes.

Since the series appeared on UPN and the WB at a time when people still thought in terms of the four networks, PBS, and "cable," it was not an easy show for viewers to find, sort of like Conan O'Brien's talk show on TBS. Even if people know about it, they don't necessarily remember to check that channel at the right time. The same problem tormented those who cared about the ratings of *Gilmore Girls* and *Buffy the Vampire Slayer*; critically praised with a rabid fan base, much of its audience didn't see the show until they got it on DVD sets.

As awareness of *Veronica Mars* grew, it attracted the attention of viewers outside the targeted demographic of the teen female audience. Many of the viewers who found the series on the DVD sets were older readers of female detective novels like those by Sue Grafton (whose Kinsey Millhone worked in a fictional city similar to Neptune), and Patricia Caldwell's Kay Scarpetta mysteries. The readers watched and appreciated the first two seasons as a pair of detective novels.

Veronica Mars was marketed as a network teen show, airing on one of the fringe networks. It also told one of the most novelistic detective stories on television while still satisfying the network demand for episodes that would satisfy in isolation. Network executives ended up killing the show by insisting on the removal of the long-arc story that made it unique.

The show was canceled amid much talk of continuing Veronica Mars's story in film, either wrapping up the story in Neptune or, following Thomas's proposal for the fourth season that never happened, picking up the story three years later with Veronica going through FBI training. Neither film happened. Rob Thomas was also rumored to have come to a deal with DC's Vertigo comics to continue the *Veronica Mars* story, but that also never came to pass, possibly because the network blocked the deal.

Terriers, the 2010 FX series by Shawn Ryan, creator of *The Shield* (2002–2008), came the closest to repeating the critical success of *Veronica Mars*, with a novelistic detective mystery driven by a focus on characters while also providing freestanding episodes that advanced the big story. While critically popular, the show appealed to uncoveted older viewers and detective novel readers, not the prime demographic of teen consumers. It was cancelled after its single 13-episode season.

The series told the story of an unlicensed private detective, Hank Dolworth (Donal Logue), in Ocean City (like Neptune, a fictional version of San Diego). Dolworth is trying to solve the murder of a homeless friend, which leads him to more murders and, eventually, a multi-billion-dollar corruption scandal. Divorced from an ex-wife he loves, partnered with an ex-thief (Michael Raymond-James) with problems of his own, they stumble toward an eventual solution and something approximating justice. In binge viewing, the series is a single well-written, character-driven, soft-boiled detective novel, destined to be appreciated more on a DVD set than it was on broadcast television.

Though very different shows, one of the shared lessons of *Gilmore Girls, Veronica Mars,* and *Terriers* is that, even though American television is corporately created, written by a room full of writers under guidance from executives at the production company and the network that airs the show, there are still instances in commercial television where a single authorial voice can stand out in the crowd and reach critical, if not commercial, success.

Another lesson is that network pressure can force changes to make the show more accessible and still lose the audience.

A third lesson is that a great DVD novel might not make for a successful television show.

House: The Premium Blend

House found the balance.

While *House* continues to break unwritten rules of network television shows, it successfully tells strong freestanding episodic shows while also telling multi-episode arcs and full-season stories. It successfully blends several television genres. More important in terms of influence, in 2008 and again in 2011, it was named the most watched television show in the world.[4]

House starts from a soap opera base, the form that developed into the medical melodrama that had peaked with *ER* and was largely played out in 2004 when *House* premiered. *House* also absorbed the disguised anthology show, with one medical mystery each week allowing for a freestanding story that could be tragic, comic, inspiring, or sappy, depending upon the week.

Like *Twin Peaks* (1990–1991), it self-awarely mocks its genre television roots while also embracing them in many storylines, the mockery most clear in the medical soap opera, *Prescription: Passion*, that House insists on watching for several seasons, neglecting his patients to revel in the fantasy of melodramatic doctors able to cure anything, before playing the same role in that week's episode. In the fourth season, episode 14 ("Living the Dream"), House finally invades the set of that same soap opera to cure an actor of a disease no one else believes that he has, so that House can go back to enjoying the show.

Perhaps most important, like *Twin Peaks, Gilmore Girls,* and *Veronica Mars, House* combines elements of different genres without sanding off all the rough edges, creating interesting dissonances and room for the creators to work with and against the conventions of different narrative formulae. There are more genres at play in the series than might appear at first glance. The key elements of the series are the hospital-based soap opera, the very specific form of the classical mystery embodied in the Sherlock Holmes stories, and the acerbic British situation comedy.

The creators made something new of the medical soap by adding elements lifted directly from Sherlock Holmes. Once noticed, House = Home(s) = Holmes, and oncologist Wilson as the enabling companion Watson becomes obvious. The drug-addicted, socially challenged genius who can solve cases without leaving his office through pure ratiocination is classic Holmes. While never explicitly stated in the Arthur Conan Doyle originals, House's cynical motto of "Everybody Lies" is certainly in line with Holmes's view of humanity.

To balance the weekly medical mystery, the show also presents a soap-ish storyline of personal crises for House or one of his colleagues that might be a multi-episode arc or last the whole season, but almost every episode provides an emotional closure, generally thematically related to the solution to the medical mystery.

Then the show offers comedy. While *House* probably would not meet most definitions of dramedy, almost every episode of the show has a comic throughline as well, often in the form of pranks. House and Wilson may be drugging each other, or competing to see who can keep a live chicken in the hospital the longest without detection, or House might reveal Wilson's past as an inadvertent porn star, or House might switch blood samples to make his interns think he has an STD. The pranks begin in the episode and pay off by the end, adding another level of closure to the weekly episode.

The result is a redundancy of episodic conclusions that makes for successful network television while distracting viewers from the long arc story that is also being advanced. Even though U.S. ratings dropped in the last years, these multiple elements were probably responsible for the show's ability to hang onto viewers long after the very formulaic "three wrong guesses, then a solution" structure became the unavoidable cliché of the central medical mystery of the week.

The creators of *House* know their television clichés and aren't afraid to use them or to work against them. When they had commercial success even after burning through formulaic stories like the "powerful antagonist" storyline and the "old girlfriend returns" storyline and the "sexual tension with a coworker" storyline, they pushed the series further. In the seventh season, they finally put House and Cuddy in the romantic relationship that, like Mulder and Scully on *The X-Files*, any American television viewer had expected since Season One. But the storyline gave very little of the happy romance network television has trained viewers to expect. Instead the long-form arc went "realistic," and then tragic.

In the fourth season, the show threw out their popular cast of "cottages" ("little Houses") and brought in a new set of doctors to compete to be House's assistants in their own version of another popular television genre, the "reality" competition. The long-season arc included the weekly closer of another "contestant" getting "voted off the island" or not getting a rose—*Survivor* (2000–) and *The Bachelor* (2002–) conventions, respectively. Viewers could tune in each week and watch the contestants struggle with a designated challenge and, at the end, House always had a twist of a reason for why that week's loser was cast away.

House's premise—and the title of the pilot—"Everybody Lies," was a direct statement of the shift that had taken place in network television. From the optimistic belief that everyone is basically honest to the cynical assumption that everyone is a liar. The cynicism spoke to the post-network sensibilities of the target demographics.

In Season Three, House meets his match in a police officer, played by David Morse, who decided to use his position to punish House, just as House used his position to punish the police officer. Each episode builds on the alternating, and escalating, unreasonableness of both men as House is arrested, his Vicodin stashes are seized, his interns' assets are frozen, Wilson is threatened with jail, House ends up in treatment and appears to kicks his addiction, walks rudely out of his court hearing to deal with a case, and only avoids jail because Cuddy lies for him. At the end of the arc of stories, rather than having learned a valuable lesson about humility or the importance of friends, House brags to Wilson that he faked his rehab and is still an addict.

This is not traditional network writing, and the extended story arc—I've left out the episodic medical mysteries because they aren't important to the long arc—gives enough time to force the viewer to change their evaluation of each character's motives and rightness several times. While, in single-episode viewing, the final revelation is funny and a triumph of House's wits, when the episode is watched as the last chapter of an extended story, it becomes a slap in the face of the viewer that violates the unwritten rule of happy endings. In multi-episode viewing, the story arc becomes the tragedy of a brilliant man who learned nothing, bragging about successfully remaining addicted while badly hurting the friend who has already suffered much to try to help House.

House: The Unlikable Hero

Another significant element to the show is the likeably unlikable lead character, Dr. Gregory House. Hugh Laurie has often mentioned that, going into the audition, he thought Dr. House was the supporting character to the more likeable Dr. Wilson.

In an innovation that was unsuccessfully copied more than it was understood, the show takes the misanthropic Holmesian diagnostician into the realm of selfishness, a hooker-frequenting self-destructive drug addict who is arrogant and verbally abusive to anyone around him. In sum, he's an admirable jerk, admirable because of his genius, not his personality. Over the years, there is some softening of the character, but the traits remain dormant, only to flare up without warning. In many ways, House is a more unattractive character than *The Shield*'s Vic Mackey or Walter White of *Breaking Bad* (2008–2012), two criminal leading men who gain our identification in spite of their crimes.

What shows like *Lie to Me* (2009–2011), *Canterbury's Law* (2008), *Shark* (2006–2008), and others that attempted to copy the abrasive genius in an old genre formula generally failed to pick up on was that the abrasion was actually the successful result of incompatible genres rubbing against each other.

It is more obvious in the early seasons that the underlying structure of *House* is that there is a medical melodrama and everyone in it acts and talks like the characters in a medical melodrama.

Everyone except House.

The character of Gregory House is not a character from a medical melodrama. He belongs in a very specific type of British situation comedy that doesn't succeed in the United States because American audiences tend to prefer situation comedies with likeable characters. By contrast, horrible characters can flourish in British comedies like Alf Garnett, the old cockney bigot in *Till Death We Do Part* (1965–1975) who was significantly softened when adapted into Archie Bunker on Norman Lear's *All in the Family* (1968–1979), Basil Fawlty in *Fawlty Towers* (1975–1979), or even more like Rowan Atkinson's eponymous character in *Blackadder* (1983–1989). Like Edmund Blackadder, the character of Gregory House is rude and selfish and says what he thinks as much toward the audience as to any character actually in the show.

In the early seasons, Dr. House is in a British sitcom while everyone else is in a medical melodrama. It works because no one appears to notice that they exist in alternate universes, and because Hugh Laurie—who came out of British comedy, including playing opposite Atkinson in two series of *Blackadder*—even with an American accent, has the droll delivery to hide the dislocation between his reality and that of the rest of the show.

Gilmore Girls functioned in a similar way, with the screwball dames speaking a different language (and much more of it) than anyone around them. Veronica Mars often speaks—or thinks—more like Philip Marlowe than a teen girl and the dissonance creates an energizing spark in familiar scenarios. *Twin Peaks* did a similar, though less obvious, blending, as Agent Cooper belongs to the 1960s series, *The FBI* (1965–1974) or maybe *Dragnet* (1951–1959), a very different reality from the soap opera that is the town of Twin Peaks. Again, the roughness of the fit is where the interesting aesthetic experiences occurred.

This dissonance wears down as *House* proceeds; other characters begin to respond to House in kind. But that element of the character standing outside the show is key to the resolution of the tension between the episodic and the seasonal stories.

The acerbic comedy of *House* adds to the episodic focus of the story, but the dislocation works much better in extended viewing from the DVD sets. When the episodes are watched immersively, the medical mysteries become unimportant. What moves to the forefront is the character of House and his individual journey of that season. His snarky asides that get laughs in the present of the episode are more revealing of his feelings in context and often serve as a deflection from his actual concerns.

House goes through significant transformations in a season of *House* that are largely obscured in single-episode viewing. The series arc is the story of a brilliant man in pain who gradually tries to open himself up to connection with others and repeatedly fails, falls back, and tries again. As the series moves toward its conclusion, it's hard to say whether the series will give viewers a happy ending or leave House in isolation and pain, but with greater knowledge.

The long arcs of *House* tend to be season-long. They follow the integration of House's team, the legal and professional consequences of House's addiction, the personal and psychological consequences of his addiction, and the failed romantic relationship with Dr. Cuddy.

The commercial success of the series has given the creators the room to take chances in their writing. In terms of long-arc network writing, the fifth season may be the most noteworthy. In that season, viewers saw House going through increasing drug-related personality problems and difficulties interacting with people. However, it was not until the end of the season that it was revealed that several of those events—including his sleeping with Dr. Cuddy—were hallucinations resulting from his Vicodin abuse. The truth of the individual episodes/short stories was revealed to be false at the end of the season and it was the last straw, leading House to institutionalize himself to deal with his addiction.

It may seem minor, but this is at least as dramatic an undermining of the unwritten rules of episodic television as anything that David Simon has done with *Treme* (2010–). With both series, there's a need to see several episodes to understand what is happening in the show but here the series, *House*, deceived the audience between episodes by presenting hallucination as reality without any of the conventional television nudges or winks that they've been trained to expect.

By comparison, the seventh season, when House and Cuddy tragically attempt to have a normal romantic relationship, seems less of a clear break from traditional television writing when viewed in single episodes. The comic elements and medical subplots—and the very special guest appearances by Candice Bergen as Cuddy's mother—obscure what is obvious in DVD set viewing. Then the distractions fall by the wayside and the viewer sees a flawed relationship fall apart, the tragedy for Cuddy and House as they both want to be together but cannot, and how badly it hurts both of them when Cuddy is forced to make the decision to end the relationship because she can't trust House, and how House's response finally forecloses on any chance of a happy ending.

The depressing material is all there in the single episodes, but it is only in cumulative viewing that it becomes the most significant storyline as the distractions fall away. Besides contradicting another of the unwritten rules of television (love should at least be happy if it can't conquer all, and "really trying" should make everything okay), the season demonstrates, again, the way that the viewing of the season as a DVD set can reveal a very different story than weekly viewing of single episodes.

House is probably not the best show on television (I have no idea what that would be), but it is the network show that has best balanced the commercial need for episodic stories with the novelistic exploration of the season-long arc. The series explores stories of dramatic transformation that upend the unwritten rules of network television, while still entertaining the network's single-episode audience.

House has been an exemplar of how a network television show can be both a series of short stories, while also exploring more complicated and disturbing themes for the novelistic reading of the same episodes on the DVD set, where the episodes serve as the chapters of a more focused viewing experience. The fact that novelistic values that come through when watching the episodes as a series of chapters does not seem to interfere with the entertainment provided by the single episodes.

Chapter Six

Men Behaving Badly on Cable

In Which the Male-Centered Soap Opera is Seen in the Early Success of HBO, HBO's Use of Network Formulae, Shakespearean and Chekovian Television are Considered from *Oz* to *Breaking Bad*, along with the Unexpressed Emotions of *Mad Men* and *The Good Wife*

The story is simple: a man lives in prison and he dies. How he dies, that's easy. The who and the why is the complex part. The human part. The only part worth knowing.

—Augustus Hill, *Oz*[1]

HBO gave the impetus to a new sort of television viewing. Even ignoring its programming, HBO is the place to start for discussing the effects of cable on storytelling. Not for anything specific in the programs, but for the memetic advertising campaign that taught Americans that "It's not TV, it's HBO."

Forget NBC's "Appointment Television." "Television" is inferior. HBO is "quality." "Different" storytelling is branded as inevitably "better" storytelling. HBO's partisans among viewers, critics, and academics have absorbed and expanded the use of the term "quality" but, much like the roughly contemporary "compassionate conservative," the term's utility depends upon its ambiguity. Quality means very different things to different people.

One of the broadest assumptions of traditional American television is that a show without high ratings is not a good show. As an entertainment medium that pays for itself through advertising sales, that's a practical definition but it has the obvious drawbacks of limiting experimentation. HBO's marketing reversed that logic and established the idea that the shows that most people *weren't* watching—and *couldn't* watch—were the best shows, turning their limited audience at the end of the 20th century into an exclusive club that offered something the average television viewer couldn't get. That brilliant piece of branding created a large space for HBO, and then other premium—and even basic—cable networks, to produce shows that network television wouldn't run. If network wouldn't run them, they must be better.

If you didn't get HBO shows—in either sense of that verb—there was something wrong with you, not the shows.

As the slogan suggests, HBO programs were sold as "not network" programming. But examining the shows suggests that, while the *content* was not network—more explicit language, sex, violence, and "adult themes," it took a while for the *form* of storytelling to stop following the network television model. While they became more explicit, more R-rated, the writing of *Oz* (1997–2003), *The Sopranos* (1999–2007), and *Six Feet Under* (2001–2005), the structure of exposition, payoff, and closure remained the same. Content changed at first, but not form. Only later did the story structures change. While *The Sopranos* was the big popular breakthrough for premium cable, *The Larry Sanders Show* (1992–1998), with its blend of fake talk show and behind the scenes documentary-esque workplace comedy, was a far greater challenge to network television formulas.

Network television gave viewers escapist dramas about easy heroes in *Starsky and Hutch* (1975–1979), *TJ Hooker* (1982–1986), and *Quantum Leap* (1989–1993). They drove well, looked good, talked smooth, and were fantasy ideals. But writers were beginning to respond to the antiheroes of Hollywood film. *Miami Vice* (1984–1990) blended film noir into their cop drama and gave viewers heroes who knew they were fighting against overwhelming corruption but did it anyway. Network was seeing a gradual transition from the cowboy hero's "a man's got to do what a man's got to do" to the existential heroism of Camus's Sisyphus, who recognizes the meaninglessness of his labors but, by choosing to continue, becomes heroic anyway.

HBO, supported by subscription and targeting a narrower audience of adults—and adult males—was ready to push further and faster than network, allowing more difficulty in stories about failure and powerlessness, acknowledging the limits of one's dreams.

Oz: The Male Soap Opera

Oz is important to the development of cable storytelling because it proved a successful hit while dealing with characters, language, events, and subject matter that had no place on network. At the same time, the form of the stories very much fit the network model. The structure of storyline payoff, the closure, and the exposition are all network. There aren't as many "feel good" moments but "feel bad" moments push similar emotional buttons in the tradition of the network soap opera.

Oz begins very much as the male version of *Prisoner* (1979–1986), an Australian "women behind bars" soap opera syndicated in the United States as *Prisoner: Cellblock H.* Besides the change in gender, *Oz*'s two main innovations from that source were that it was shot on video in a documentary style on a open-plan set that underscores the environmental "realism" of what was shown, even as it had a theatrical sensibility, and the fact that show overtly addressed one social issue in each episode, with other timely issues informing the longform soap opera arcs. Some arcs were season-dependent, others seemed shorter but, in the tradition of soap operas, most of them burbled along under the surface, disappearing only to break through from time to time, until finally wrapping everything up in the last episode of the series when a mass poisoning in the prison led to the surviving prisoners—and, simultaneously, the entire show—being loaded onto buses and the lights being turned out.

Over the years, the "men behind bars" soap opera never disappeared from *Oz*, neither did the overt treatment of "issues," but there was an attempt to expand from melodrama into a Shakespearean sense of tragedy that became overt in the final season as the inmates work on a production of *Macbeth*.

What separated *Oz* from the network fare was the explicit addressing of sex, drugs, rape, violence, racism, and the inability of the authorities to do anything but maintain the appearance of control. The ideology of the show shifted from the guarded optimism of a *Hill Street Blues* (1981–1987) to the sort of humorless fatalism that had been seeping into shows like *Wiseguy* (1987–1990) and *Miami Vice*. The testosterone-soaked environment of the show was present in *Hill Street Blues*, where violence might break out at any moment, and women needed to find a way to adjust to a male environment of honor and revenge, but it was the world of the male soap opera taken to a Hobbesian extreme.

Over the years, the show taught its viewers about the banality and humanity of criminals, rounding them out into more complex figures, but the traditional narrative form constrained the stories. There was the need to match the expected rhythm of beats of television narrative and closure. Similarly, the moral choices were clearer and viewers got episodic resolutions, even if they were unhappy closures.

Everything in *Oz* is serious and unironic at a time that irony was seeping into most of American discourse, but especially television drama. *Buffy the Vampire Slayer* (1997–2003) and *Ally McBeal* (1997–2002) represent the tone of the era, perhaps not surprisingly, male-written depictions of female empowerment. Humor deflects and releases tension and *Oz* was fueled by its tension.

The series is narrated by Augustus Hill, a wheelchair-bound inmate and a poet who serves as the Stage Manager of HBO's bleak *Our Town*. Hill soliloquizes, even breaking into free verse, dragging viewers to an interpretive level that explains greater truths but also reminds viewers of the artifice of the show's extreme "realism."

Hill's narration breaks the fourth wall to introduce characters and highlight the messages of that week's episode, existing both within the narrative and outside it, highlighting the theatricality of what we watch. At the same time, the narration provides comfort, reminding viewers that, regardless of how disturbing the show might get, they will have everything explained to them in simple form. Hill's mediation provides the relief in a stark and humorless show.

Oz ended after five seasons with a series of deaths meant to resonate with the killings at the end of the production of *Macbeth* that the prisoners were staging. Among the fatalities was the white supremacist whose death some viewers had waited for since the very first episode of the series. The viewers' possible satisfaction was removed by the fact that his killer was tricked into killing the man, turning his revenge into evidence of his own powerlessness.

Through its poet narrator, *Oz* provides a sort of theatrical self-referentiality that jars with other television of the day, deliberate artificiality juxtaposed with social realist documentary style and subject matter. In that, the series might be considered a television version of the prison poetry performed at regular intervals by inmates within the series, claiming veracity through its gritty realism and artistry through stylized allusions.

The Sopranos as Literature

David Simon once suggested in passing that *The Sopranos* (1999–2007) and *Deadwood* (2004–2006) took their inspiration from Shakespearean tragedy, while *The Wire* (2002–2008) drew from Greek tragedy.[2] There is reason to agree with the statement as something more than the idea that Shakespeare is the gold standard of dramatic narrative. *The Sopranos* tells the story of a king whose crown never sits securely upon his head, and the prices he must pay to keep it there. The first season finds Tony expected to take over the family but his Uncle Junior holding on to power and plotting with Tony's mother to kill him.

If one insists on the "fatal flaw" approach to Shakespearean tragedy, Tony Soprano is intelligent or sensitive enough to be aware that there is something wrong in his life even as he is unwilling to do anything to change his situation. For Shakespeare, it would have been called "conscience," although Tony would probably reject the term.

The Sopranos does follow Shakespeare in performing variations on this theme: besides Tony, his nephew Christopher wants something more out of his life but increasingly finds himself unable to move into film work or just get away, and turns to drugs to deaden his awareness. Tony's wife, Carmella, rings another variation on the theme as she runs up against the limits of her privilege, Tony's philandering, but also his brutal nature. In the sixth season, Vito Spatafore, a member of Soprano's work family, is able to recognize his problems (he is a gay man in a deeply homophobic environment) and, unlike everyone else, acts to change it. He leaves the mob to live his true life and is happy. Tony comes near to envying the man before having him killed to protect the Soprano family honor.

Perhaps the limits of Tony as a Shakespearean hero lie in the fact that Tony never realizes the tragedy of his situation, he just continues on. Which may suggest that Arthur Miller's *Death of a Salesman* is at least as relevant to what *The Sopranos* is doing.

The Sopranos is a post–World War II tragedy. Perhaps post-Vietnam would be more descriptive, but after World War II, the idea of the existence of Great Men, and the tragedies of Great Men, came to be out of step with Americans' experience of their world. Instead Americans turned to the great tragedy of the Average Man, best exemplified by Arthur Miller's 1949 play, *Death of a Salesman*. Willie Loman is forced to face his own inadequacy, his lack of a future for himself or his sons, and his lack of a past that he can remember with honesty. Loman is left with a dream of himself that no longer works. Even his last attempt to turn his failed life into a small triumph turns pointless.

Novelistically, Robert Penn Warren straddled the Great Man/Average Man divide in his 1946 novel, *All the King's Men*, where an average man tells the story of a possibly great one, both of whom fail to live up to their own ideals and their ideas of themselves and each other. In the process of telling the story of these characters, a broader story about the corruption of America's dream is more broadly revealed.

Cinematically, the Average Man tragedy of screenwriter Robert Towne's unfinished *Chinatown* trilogy—with *Chinatown* directed by Roman Polanski—examined the sweep of Los Angeles' history and decline from the perspective of the iconic private detective from the 1940s through the 1960s. Standing alone, Polanski's *Chinatown* film established the major themes of the Average Man who sees himself as great and ends up destroying that which he set out to protect because he trusted himself too much and could not imagine how corrupt the world could be.

When Tony Soprano explains to his therapist that "it's good to be in something from the ground floor. I came too late for that. I know. But lately I'm gettin' the feelin' I came in at the end. The best is over,"[3] he is expressing the great tragedy of the Average Man. Dr. Melfi makes the point explicit when she responds that "Many Americans feel that way."[4]

The Sopranos and Closure

The Sopranos displayed many things on HBO that were "not television"—the colorful obscenities, the contradictions of standard values, the occasional nudity, the objectively unlikable, immoral central character, and the wallowing in moral ambiguities. The series also offered a more cinematic use of the frame, a more cinematic writing style, allegorical passages, long arcs, and meandering plotting.

On the other hand, it also conformed to television in meaningful ways—however immoral, Tony Soprano became likeable because of his self-awareness and being better at his job than anyone around him. The audience's surrogate continued to triumph over adversity, even if the specifics took away some of the gratification for viewers, the narrative form kept it comforting. The story was still a mob series with recognizable rules and the stimulating twists of killings and betrayal, suspense, but always offering closure.

Except once. The "Pine Barrens" episode from the third season began with Paulie and Chris trying to bury a dead Russian in the New Jersey Pine Barrens. The Russian turns out not to be dead and, in trying to catch him, the two get lost and have to make their way back to civilization through freezing winter in the natural world where they clearly do not belong. While the episode won an Emmy and great critical acclaim, fans continue to harass Chase about what happened to the Russian. Which says something about the amount of closure viewers expect from a television show, and felt they received from *The Sopranos*.

Every *other* episode gave the viewer a sense of closure using humor and or violence to ensure that the episode paid off on the viewer's investment, even if the long arc remained unresolved. A moralist could watch the show and see most of the characters suffering if they wanted that validation. An immoralist could enjoy the things that Tony and his crew got away with.

The greatest formal accomplishment of the series may be the way it continued to honor the most basic demand of American television—characters don't change—but turned that into a central element of the series' tragedy.

The running theme of the series, a sort of existential joke woven through its fabric, is that Tony Soprano starts the series by going to a therapist because he is having anxiety attacks that are interfering with his work. Viewers are led to expect Tony to change, both because of the classical narrative structures of transformation, but also because of the demand for happy endings in television. After years of therapy and repeated awakenings to his need to change his life, he never does.

For all the killings, double dealing, family crisis, and duck incidents in the swimming pools, Tony Soprano is a classic television character. He doesn't change.

The writers of *I Love Lucy* (1951–1957) would be proud. In every episode of *The Sopranos*, Tony is the same selfish thug as every other. Even taking hallucinogenic drugs and getting the big picture, he is still Tony. The writing achievement of the show is that, within that unchanging character, David Chase found wonderful variations and revelations.

But Tony doesn't change.

And that television convention, so common it's generally unworthy of note, is what makes Tony, and *The Sopranos*, a genuine tragedy. Because the arc of the entire series is that we start with a mobster who is having a breakdown, who is forced by it to explore who he is and what he wants, to move toward self-awareness.

Instead, Soprano increases his self-awareness and rejects it. He uses his psychotherapy and medications to become a better mobster. But he refuses to change. He is the same man in the last moments of the series that he was in the first episode.

This is the ironic purity that explains the end of the series. At the end of the last episode, Tony Soprano and his family gather in a restaurant that seems filled with potential killers, either alluding to people wanting revenge from previous seasons or providing shots of characters the series has taught us to be wary of, with even an invocation of *The Godfather* (1972), as one character heads to the bathroom like Pacino's Michael Corleone before assassinating a mob boss. In this heightened state of allusive anxiety, Meadow almost gets hit by a car while jaywalking across the street and then . . .

. . . nothing.

The screen goes black and that's it, because this is an American television show. And American television shows go on forever. The lead characters cannot change. And killing Tony would be a permanent change.

What is normal for network television, the unchanging character in an unchanging life, is depicted as a tragedy of banality in *The Sopranos*. Amusingly, the same point is made in the ending of *Seinfeld* (1990–1998) but without the added emotional weight. The last lines of the last episode repeat the first lines of the first episode, a discussion between George and Jerry about wrongly placed shirt buttons. After nine years, *Seinfeld* also went nowhere.

The added irony of the ending is that viewers were outraged because the show left the characters forever unchanged. *The Sopranos* left the viewer in the uncomfortable position of an observer confronted by Schrödinger's cat. Tony is forever sealed in the black box of television, neither dead or alive, until someone makes a sequel.

The Sopranos ending frustrated many of its viewers because they had bought into the slogan of HBO—that it's not TV, it's HBO. *The Sopranos'* last act was to remind them that it was still television after all.

Time has proved this out. *The Sopranos* works quite well in syndication. With a couple of cuts—necessary anyway to shorten the episodes to allow for commercials—and a few bleeps or dubs, it fits right in with all the other *Bones* (2005–), *NCIS* (2003–), and *Law & Order* (1990–2010) reruns on basic cable.

In part because of its humor.

The Sopranos as Comedy

With its popularity and HBO's support with reruns and DVD sets, viewers could be expected to see every episode of the show in sequence.

As someone who watched *The Sopranos* in five- or six-hour binges, I mainly remembered the tragedy of Tony Soprano and the people trapped in his life with him. There was always humor, but it was a dramatic show. But a funny thing happened when I started watching random syndicated episodes out of sequence to research this book.

I discovered that *The Sopranos* was a comedy.

The removal of nudity and obscenity and a few other minor details for mainstream broadcast didn't dampen the tragedy. It was the act of watching *The Sopranos* in a few freestanding episodes, rather than in sequence as part of a long story.

Without the sequential episodes and the long arcs, the centrality of the family melodrama and the tragedy disappears, what is left is some unconnected episodic drama and small tragedy, but lots of comedy. In isolation, the absurd behavior of the characters—even in tragic situations—becomes comical.

There are different types of comedy. The classical dramatic definition of comedy is shorthanded as "lover's reunited." That definition doesn't apply to American television comedy, where we tend to break television into drama and sitcom, with something newer called dramedy somewhere in the middle. But the humor of *The Sopranos* is more complicated than that. It comes closest to what Andre Breton called black comedy (*humour noir*), the comedy of cynicism and death.

From its opening moments, *The Sopranos* announces itself as a comedy; a snicker-inducing shot of Tony's disorientation in the presence of a Robert Graham bronze nude in Dr. Melfi's waiting room, followed by Tony's deadpan euphemisms for his "waste management consulting" job that includes "having coffee" with a debtor, especially when juxtaposed with viewers seeing him using his nephew's new Lexus to run down a guy who owes him money with Dion singing "Don't Know Why I Love You Like I Do" on the car stereo. And then, with the broken man lying on the sidewalk beside the car, nephew Christopher can only see that a headlight's been damaged. Later Big Pussy has to explain the meaning of Luka Brazzi in *The Godfather* to film maven Christopher, who also has learned the wrong lessons about how to dispose of a body from that movie. And AJ, Tony's young son, responds to the news that his grandmother won't be at dinner with a disgusted "So what? No fuckin' ziti now?"

The comedy of *The Sopranos* is painful and ironic.

Chase has discussed the fact that his film idea that became *The Sopranos* was a comedy about the angst created for a mobster by putting his mother in a nursing home. Perhaps because *The Sopranos* premiered in the same season as *Analyze This* (Peter Tolan and Kenneth Lonegran's 1999 Harold Ramis–directed comedy that featured Robert De Niro as a mobster seeing a therapist played by Billy Crystal) most of the initial promotion for the series, and subsequent discussion of it, has downplayed or ignored how funny the show is.

Formally, the premise of *The Sopranos* is funny. The ideal of the mob films like *The Godfather* is juxtaposed with the banality of a present-day mob boss with none of the cerebral sophistication of Brando's Don Corleone. The mob that Coppola depicted as the last bastion of family honor becomes a cross between a soap opera of marital disputes and adulteries and a family sitcom about TV during dinner and grounded kids sneaking out and not doing their homework. This is comedy of deflation, the juxtaposing of a romantic ideal with a banal reality.

To once more counter the tendency to view *The Sopranos* as Shakespearean, in terms of comedy the series seems to owe more to Anton Chekov, the Russian writer of comic-tragedy or tragic-comedy plays including *Uncle Vanya*, *The Seagull*, and the *Cherry Orchard*. Just as *The Sopranos* is mainly discussed as serious drama, Chekov's plays have often been staged for their Russian tragedy without any acknowledgment of how much comedy the playwright rings out of the inertia and despair of his characters.

While they aren't as educated as most of Chekov's characters, the characters of *The Sopranos* are not stupid. Instead they are ignorant and mean (in both the modern and the broader traditional senses of the word). Tony and Christopher are intelligent enough to see what is lacking in their lives, so they offer viewers two variations on the theme of that existentialist core of the series. Tony ends up in therapy after adultery, food, and money proved not enough. Christopher escapes into drugs and the fantasy of a film career. But like their less self-aware peers, their knowledge is very specific to the world of the mob in a part of New Jersey. Visits to New York City are like traveling to a foreign country with an entirely different culture.

One steady stream of comedy arises from the mobsters' ignorance of other ways of doing things, of alternate meanings of words. Tony's "Guys today have no room for the penal experience." The other flows from the fact that most of the characters are pure id, spoiled children driven by greed and envy.

Much of the humor is in the childishness of dangerous men. Paulie, especially, is very funny in individual episodes. His greed and impulsiveness are absurd in an adult man. In the context of long arcs, Paulie is scary for exactly the same reasons; it's the episodic comedy that adds disturbing complexity to his character. Paulie's rejection of his mother when he finds out she didn't give birth to him is ridiculous in the episode, only in the context of a broader arc is it tragic.

By comparison, Christopher mugging Lauren Bacall for her goodie bag in the sixth season is a simple gag that reminds the audience that, for all his aspirations, he is still an animal.

Watch a couple episodes back to back and the comedy fades into the background as we watch the consequences of the absurd, childish behavior. But it is a comedy.

Chase elevates what was the network formula of the day. *The Sopranos* blends the male soap opera with the mobster genre. Serious drama is leavened with comedy; the wacky situations of misunderstanding so common on network television are played for comedy *and* tragedy. For all the changes, the formula is still there, just adjusted.

There is a reason that the end of *The Sopranos* is the same as the end of *Seinfeld*, a reminder that nothing ever changes in the television story form.

The Antiheroics of *The Shield*

The ending of FX's *The Shield* (2002–2008) provides an interesting contrast to *The Sopranos*. Instead of ambiguity and indeterminacy, *The Shield* gave us a moralizing finish in which detective Vic Mackey (Michael Chiklis) escapes all legal punishment for his many crimes and is instead given a job with the FBI, but he suffers a greater punishment than he could have imagined, as the man of barely contained violence is condemned to a bureaucratic desk job that, for Vic, is nothing less than a fluorescently lit cubicle hell.

The Shield, created by Shawn Ryan, was often treated as sort of a *The Sopranos Lite*, an amoral police detective in a more episode-centered storytelling on basic cable, in theory, halfway between network and HBO. Based on the Rampart Division police scandal in Los Angeles, the series followed a cop in charge of a tight group of police who have become a law unto themselves as they attempt to fight gangs and drug dealers in Los Angeles. *The Shield* told more tense and driven stories than *The Sopranos*, and its single episodes provided closure while also serving an obviously serial narrative, requiring multiple-episode viewing in a way that most other series at that point did not.

The Shield moved more toward the network model of a powerful and charismatic lead character, even as he was allowed to be morally reprehensible for long sections of the series. In television terms, he is still a great lead character because, though venal, he is successful, he solves problems, imposes his will on others, defeats those who would interfere with him, and cares about his wife and children.

The fact that the people Vic defeated were often, technically, the good guys, and he was committing crimes up to and including murder to meet his goals, added a complexity to the character, but one that calls back Sonny

Steelgrave and Mel Profitt from *Wiseguy*. The character of Vic offered the chance to explore the differences between morality, immorality, and amorality within the structure of a recognizable cop drama, but he also offered a cathartic release in his triumphs by the end of almost every episode.

Vic did immoral things—stealing, killing, lying, blackmailing—to achieve moral ends (stop drug dealers, punish criminals, protect his family). But he used the same tactics to achieve selfish or venal goals—stealing millions of dollars from drug dealers, cheating on his wife, and manipulating those who loved him. Vic could and often did rationalize what he did, but the show rarely let him off the hook.

The series continually returned to difficult questions about how immoral acts can serve moral ends and how moral standards create immoral situations, but at its heart was Vic, a bad cop who did good things by breaking the law, with complicated motives and rationalizations. Like Tony Soprano, he was good at what he did, and viewers identified with him. Perhaps in recognition of the moral quandary created by doing this on basic cable, over time Vic was increasingly moralized by emphasizing his more acceptable motives of loyalty, protecting family, desire to punish the bad guys, and by offering antagonists that viewers could see as worse than Vic.

Glenn Close and Forrest Whitaker were brought in, in later seasons, elevating the perceived quality of the series and giving Vic worthy adversaries the audience could use to forgive Vic for his excesses/trespasses, but not completely. Stacking the deck so that those who would punish Vic used immoral tactics to bring him down added layers of moral complexity but also made it easier to cheer for Vic. Like the minimized obscenity, moral lessons straddled network and premium cable standards.

To continue the Shakespearean comparison made by Simon, Vic Mackey is probably the most genuinely Shakespearean character on cable in this era of television writing, in the strict sense that Vic fits the mold of the Great Man that Shakespeare depicted—a great, though amoral, cop, a larger than life hero. Just as Macbeth and Titus Andronicus are great battle generals who fail in the political and family aspects of life, Vic Mackey is seen to be a great tactician and street soldier who gets tripped up in resolving what works with what is right. Tony Soprano, by contrast, is never a Great Man; he's an Average Man in the no longer great role of mob boss.

Wiseguy (1987–1990) had long since opened the television depiction of police acting immorally for a greater moral good, with Vinnie Terranova tormented by the struggle. Vic Mackey blew that romantic sense of honor and remorse out of the water, replacing it with a multimillion dollar hijacking of drug money for personal gain, drug dealing, and a sizable number of murders. He knows he is doing wrong but justifies his corruption by self-righteousness platitudes about honor, defending friends and protecting his family, but the series winds toward an end that is the modern equivalent of the Jacobean dénouement of a play like *Titus Andronicus*. Jettisoning by stages his friends, his family, and his honor, Vic ends up getting away clean only to have willingly consigned himself to the worst punishment he could have received.

Here again, the tragedy of the modern era comes into the drama; the great man of action is condemned to be average, working a desk job under sputtering fluorescent tubes, with no escape until retirement. It is as if Macbeth was not killed at the end of his play but instead was made to live out his life refilling the inkpots of copyist monks between performing hourly prayers.

Perhaps ironically, Mackey's punishment is the life of countless viewers of the show who had watched it for the vicarious escape from the drudgery of a desk job in the violent man of action breaking the rules and getting things done. Vic is punished by being turned from a TV hero into a member of the audience.

The Sopranos plumbed the ironic tragedy of the mob boss being one of us. *The Shield* explored the man who stands outside society as both hero and villain. The same sort of figure Clint Eastwood cut in *High Plains Drifter* (1973) alone, or *The Wild Bunch* (1969) revealed themselves to be in the most romantic of cynical setting. Unlike those films, or even the Charles Bronson vigilante film, *Death Wish* (1974), *The Shield* ends by stripping Vic of his heroic trappings and rejects the myth that led many to watch the series in the first place.

Breaking Bad: The Television Tragedy

Just to finish the slowly developing thought: *Breaking Bad* (2008–2012) is the most Shakespearean of contemporary television series in story, if not in characters.

Chemistry teacher Walter White and one of his worst students, Jesse Pinkman, are average people who become meth cookers. They are characters from the realm of Miller's salesman, not the Great Men of Shakespeare, and they live in a world where neither heroes nor villains have moral superiority for even the length of an episode, underscoring the situational morality that puts the series in contrast to the verities that hang over the work of Shakespeare.

It's hard to make any definite claim about *Breaking Bad* because, more than any other television series, it has rejected the traditions of television and continually changes the tone of the story and the nature of the characters as it develops. As its creator, Vince Gilligan has repeatedly said, the lead character changes from Mr. Chips to Scarface. The result is a series that uses the format of the DVD set to create a Shakespearean consideration of characters with the scale of transformation only appreciable by viewing a season or the whole series in compressed viewing.

Walter White (Bryan Cranston) is a man filled with justifiable rage in Season One, who becomes a meth cooker to pay his medical bills when he finds he has cancer. The season opens with White in his white briefs holding a gun and talking to a tape recorder on a desert dirt road, with police sirens getting louder and a smoking mobile home in the background.

It's a comedy, or a dramedy, depending on where you come in to the absurdity of the story of a high school chemistry teacher with cancer going into business with a stoner former student to cook the best methamphetamine in the American Southwest.

Walter is powerless in the classroom, powerless in the faculty lounge, powerless at home, powerless at his second job at a car wash, and powerless in the face of doctors telling him he has cancer and an insurance company telling him they won't cover the payments. He is powerless in the way most of his audience is, but he fights back in the way television audiences like to watch. He sets out to get the money and, in the process, takes power into his own hands.

But almost immediately he goes too far. He sets a jerk's car on fire for stealing a parking place. He breaks the leg of a kid who mocks his son. The first is funny because the audience doesn't like the jerk, but the second is scary, even more so because we see his family's reaction to what he does.

Still his wife and son love him, even if they push him around, his students ignore him, his car wash boss treats him badly, and he's got incurable cancer. So viewers identify with him. Feel sorry for him. Like him.

Part of the skill of the show's construction is that starting from that beginning, working against the television principle that we will identify with the protagonist, especially if he does things well, the show goes on to reveal Walter to be an increasingly unlikable man. Whether he reveals what was always there or whether he becomes unlikable as he feels more powerful is open for debate.

At the same time, as he's fighting cancer and deceiving his wife about where the money is coming from, the show reveals that the love of Walter's life was stolen from him by the fellow chemist who became a multimillionaire off Walter's and his shared research, when Walter was left with nothing. Besides suggesting Walter's wife Skyler was a settling second choice, the revelations make Walter all the more sympathetic.

While this summary doesn't do it justice, the first season is a very funny story about a man faced with cancer taking control of his life by becoming a criminal. The stoner former student, Jessie, provides much of the bleak comedy, for example when he tries to take a shortcut and dissolve a corpse with acid in his bathtub instead of buying the proper container, leading to the acid eating through the tub, the floor, the ceiling, and splattering human jello through the house, or when he forgets to cancel an open house to sell the property after he has set up a meth lab in the basement. Together, the duo comically attempts to rob a chemical warehouse between guard shifts. And Walter is revealed as very bad liar as he attempts to explain the new money to his wife and her DEA agent brother-in-law, neither of whom seem to notice.

But tragedy is seeping into the story as the moral consequences of his decisions begin to reveal themselves to Walter. Murders are only the most obvious examples. Still, the tone is comic, albeit blackly, because Walter is good at making meth and that gives him a bit of power to fight back against forces the audience doesn't like. It's roughly the same "justifiable immorality" comedy as the Tom Cruise film *Risky Business* (1983), where a high school boy turns his house into a brothel while his parents are gone and has sex with Rebecca DeMornay and gets into Harvard as a result.

In the second season of *Breaking Bad,* the comedy disappears. Structurally, it is still there. There are amusing reversals, scenes in which Walter is forced to lie badly yet again, awkward professional misunderstandings as Walter begins to apply the same logic he used to cook the best meth to becoming the best dealer he can be, but we watch a decent rational man methodically turn himself into a monster. In the third season, this only increases, with Walter's attempts to help Jesse revealed to be destroying what turns out to be a surprisingly sensitive boy who was there under all of the drugs.

It is hard to say how the final season will end, except badly, but the show has been laid out as a tragedy that examines the nature of evil. Walter White's choice to become Heisenberg (his professional alias, and an implicit rationalization as it invokes the uncertainty at the heart of the universe, as opposed to moral absolutes) specifically parallels Macbeth's choice to kill the king. The Scottish play gives Macbeth the excuse that his power-mad wife drove him to the murder, and the play's tragedy is revealed as Macbeth comes to understand how little he got in exchange for his choice to knowingly accept eternal damnation and the loss of everything he held dear (from his wife's sanity to his own sense of honor and his eternal soul).

Macbeth, the Great Man, makes one bad decision and then things escalate as he is forced to deal with the consequences. *Breaking Bad* is a much longer story. It allows viewers to watch Walter White, the Average Man, choose to be evil and then continue to choose to do evil things. Walter may be pushed by circumstance, but he makes a rational decision to become the villain, and then become a better villain.

This is a new depiction of evil in American popular culture, and definitely a more disturbing and complicated version of evil than has been considered on television. Walter White is an intelligent but powerless man who chooses, step by step, to break the law, to kill guilty people, to cause pain to those he says he loves, to endanger innocents, to kill innocents; each decision is made in a logical manner toward the end that he rationally decides to make himself a monster.

As Walter proceeds, the show reveals the damage his actions have caused to everyone around him, while also revealing how quickly everyone around him is also willing to abandon their own morality. By Season Four, Skyler, his very moral wife, has embraced the drug money as her own and the opportunity to use her accounting acumen to hide the profits, and she is well down a similar path to the one Walter took in Season One, taking one man's lifework away from him and causing the death of another.

By Season Four, there is still humor but it's mainly an indication of tragedy. Near the end of the first episode, Walter and Jesse wear absurdly matching Kenny Rogers T-shirts and ill-fitting white pants as Jesse eats pancakes in a Denny's by the highway. Walter's shirt still has the size tag stuck on Kenny's face.

It's a hypernormal scene that skews toward comic relief, as they are alive after having spent most of the episode expecting to be executed. But Jesse and Walter are wearing the ridiculous T-shirts because their own clothes were covered with the blood of a man killed in their place, in front of them, who they had to dispose of and clean up after. The unremoved size tag on Walter's shirt is funny until the viewer realizes it is a sign of just how disturbed the fastidious Walter still is about his near-death experience. It's also an early indication that Jesse has gone somewhere else as he enthusiastically eats the meal that Walter can't touch, confirmed when

Jesse articulates the nihilistic conclusion he has reached after a night that began with him killing an innocent man, then waiting to be killed himself before being splattered with the blood of another, before he got his Anytime Pancake Breakfast at Denny's.

The conclusion is that Jesse has decided that he is "the bad guy" so he's going to accept it.

This nominally comic scene is the moment when the viewer is told the innocent Jesse has left his humanity behind, that he is capable of making cold-blooded calculations in what he now sees as a world of mutually assured destruction. Walter's three seasons of rationalizing that he is helping Jesse are revealed as a cruel joke; Walter hasn't saved Jesse, he has doomed him.

The most interesting development in Season Four is the revelation of drug lord Gus Fring's epic backstory, his journey to the United States, building an international empire, and finally taking revenge for a murder long-cooled. In a traditional dramatic sense, Gus (Giancarlo Esposito) is the closest thing to one of the Great Men of the Shakespearean stage—a man who has done great things at great costs to himself, a leader, a builder, and a man willing to risk everything on his certainty—and he plays the antagonist to the average Walter White. It is a genuinely modern tragedy when the Great Man is brought down by the Average Man of the modern era, a man who will never achieve greatness himself and who is well on his way to ensuring his own destruction. Similarly, Skyler increasingly reveals a latent modern Lady Macbeth, capable of great acts of evil but only when she does not have to acknowledge them as such.

The accomplishment of *Breaking Bad* is that it has successfully built from a first season of black comedy, to a second season crime suspense story as Walter builds his empire, with the third season turning darker as Walter and Jessie go to work for Gus, and a fourth season that was unremittingly bleak and only relieved by gratifying murders and revenge at too great a price. Going into its final season, *Breaking Bad* has been able to take its audience along on a painful and depressing journey that equals the bleakness of some of the lowest points of *Six Feet Under* (2001–2005), but with more paranoia added to the mix.

Any single episode of *Breaking Bad* is structured to provide an emotional payoff but it is clearly constructed for viewers to see every episode in sequence. The series serves the commercial imperative—it's compelling entertainment with subject matter appealing to young male demographics, but it also examines morality with a complexity we don't expect of American television.

In viewing each season as a novel, or a section of a five-part novel, the audience sees tidal shifts as characters change and roles reverse. One of the most disturbing elements of the series is that the drug-dealing former student, Jesse Pinkman, who seemed dangerous but addled in the beginning, increasingly is revealed not as Walter tempter but as Walter's victim. At the end of Season Four, Walter's actions remind viewers that he will do anything to Jesse to get what Walter wants.

According to the traditional rules of television, *Breaking Bad* simply can't exist. Its subject matter, its complexity, it's length, its willingness to embrace the structure of a Shakespearean tragedy, and all of the transformation of characters over time in response to events that is required by drama in the nontelevision media, moving its audience from comedy to tragedy, and even its exploration of personal responsibility at a time—and in a crime story setting—that largely rejects the idea, make the series an exceptional example of television and a series that will be re-understood when it is only available on DVD sets.

The Male Soap on Cable

Having repeatedly returned to David Simon's throwaway comment about Shakespeare, it would be irresponsible not to mention the one series that started out as an explicitly Shakespearean series, *Sons of Anarchy* (2008–). After beginning as an overt remake of Hamlet in a motorcycle club—club president Clay Morrow (Ron Perlman) having killed his predecessor and married his widow (Katey Segal), is raising as his own son, Jax (Charlie Hunnan), the murdered president's son who doesn't know whether to take over or leave—*Sons of Anarchy* soon dropped the Shakespeare and moved more into season-long stories with episodic closure of the club in conflict with external forces, including white supremacists moving into town, the IRA fringe gun runners who kidnapped Jax's illegitimate son and the blackmailing of Jax's mother by a corrupt FBI agent who framed her for murder. The series continues to deal with the guilt of the mother and club president, as the son and his civilian girlfriend make plans to leave the Sons of Anarchy motorcycle club, but the themes of betrayal and the personal struggle of the prince have returned without the initial references.

All of the series discussed so far have followed the example of the male soap opera, asking us to identify with the suffering of the male in his workplace—be it a prison, the back room of a strip club, a meth lab, or a motorcycle club office. *Breaking Bad* stands out because it led viewers into the story through identifying with a protagonist's unfair suffering only to repeatedly force the audience to question its identification.

Six Feet Under presents the male soap opera in its unblended—and un-prisoned—form. Following two brothers taking over the family funeral business after the death of their father, the workplace is literally the home and, while strong female characters balanced the men to some extent, the brothers were the heart of the show, with the third "brother," Frederico Diaz, who actually does most of the body work and feels increasingly taken for granted as the series goes along. There is the vestigial remnant of the disguised anthology show in the opening death and the episode's featured funeral, but rarely do the mourners become central to the story. Instead, the show tells the story of people in the business of dealing with death for the living, who cannot make sense of death or life for themselves.

The most interesting stylistic aspect of show, beyond increased realism of content for a soap, was the overtly theatrical device of the brothers regularly talking to ghosts to externalize the emotions they are going through. It's been suggested that any devices that take us into the subjective reality of a television character are derived from *The Singing Detective*, so this might be another example, along with Tony Soprano's epic dreams of forewarning on *The Sopranos*.

HBO also offers the most distilled essence of the soap opera in the series, *In Treatment* (2008–2011). The series, starring Gabriel Byrne as a therapist, presents the weekly therapy sessions of three or four patients, each with their own weekly session on the same night each week, and then observe the therapist's own session with his analyst as well. *In Treatment*, which began as a translation of the Israeli series, *Be Tipul* (2005 & 2008), has essentially removed the action and distilled the soap opera to its essence, someone talking about how they are feeling while someone else listens and tries to help them. While more extreme than other television drama, it speaks the language of television viewers.

In Treatment's three seasons provided a viewing-changing variation on longform television. With the same character on the same night, viewers can watch one night a week and follow that patient's story. Or they could watch a couple, or all, including watching at the end of the week when viewers got to hear what the therapist was thinking and feeling. The narrative possibilities are both vertical and horizontal.

It could be said that AMC's *Mad Men* took the soap opera in the opposite direction from *In Treatment*. Another male soap opera, *Mad Men*, asks us to empathize with the suffering of the post–Korean War male, with some access to the suffering of the women as well. In contrast to the wash of information provided by each *In Treatment* patient, *Mad Men* is built on concealed emotions rarely expressed.

The stories of advertising men (and women connected to them) after the Korean War are fairly conventional stories of betrayal and disappointment within the genre, even taking into account the fact that Don Draper started his life over by taking the identity of a dead man to get home faster from the war and to become free of his past.

But while the stories are conventional, the way they are told is something that needs to appear on cable—or on DVD—for the audience to understand how to watch it. The series takes its narrative structure from the repressed feelings and uncertainties that are associated with America in the 1950s and 1960s. Where *In Treatment* presents people of the new century spilling their guts, the characters on *Mad Men* hold everything in, and don't know how to express what they can't acknowledge anyway.

Mad Men is not a flowing narrative. The series has a season-long arc and smaller multi-episode arcs, as well as closed single episode stories, but the real narrative focus is smaller than even that. An episode of *Mad Men* is a series of scenes that function as zen koans. The scenes don't advance a story as much as they reveal character and then end, leaving the viewer to contemplate what they have just seen. It could be compared to *The Killing* (2011–) or *Rubicon* (2010), two other series that use a slow, opaque narrative. All three series use pacing and silence to force the viewers to figure out what the mystery is, and anticipate what will happen next.

The difference is that, technically, *Mad Men* is not a mystery. But it is, in the sense that the conceit of the show is that people don't know what they want and don't know how to say it if they do, so nothing that is said or felt is quite true to the speaker.

Mad Men asks viewers to watch a scene and then contemplate what just happened and what it means. Sometimes what it means in terms of the story but more often than not, what it means about the person and what it means about existence.

Mad Men presents a stylized version of the American past to comment on the present. The *Mad Men* version of 1950s/1960s culture seems much like present culture, but then suddenly turns out to be different from what we expect. Or it seems different but suddenly becomes all too familiar. These little touches give the aesthetic and emotional power to the series. Don Draper looks like the hero and is presented to viewers with all the television trappings of heroism; viewers follow him and spend much of their viewing time with him. He's self-created, he suffers, but he's also narcissistic, an adulterer and a bad husband. The series forces viewers to ask if his jerkiness is a product of his time or if he's just a jerk. And just as viewers begin to decide he is just a jerk, Draper shows sensitivity or he starts to try to turn his life around. Or vice versa. The tension of the show is that created between the present and our idealized positive and negative views of the past.

The meditative nature of the series survives the commercial breaks of basic cable, in fact the commercials actually enhance the narrative because they become a more extreme version of a black out. The audience has to wait and ponder what they've seen. *Mad Men* may be the one show that actually suffers on DVD because the scenes proceed one from another without a break in between for contemplation. But there is always the pause button.

This contemplative effect is enhanced by a very unusual writing style for American television. The characters on *Mad Men* only occasionally explain what they are feeling, and generally don't do it very well. Thematically, this reinforces the mood of everyone's lives being constricted and guarded, or sends the message that the characters are unable to articulate what is wanting in their lives. However, formally, it creates an unusual experience for viewers because American television characters are always talking about their feelings. Traditionally, network television has been about making sure that the audience always knows what is going on, creating comfort at not having to think to figure out what is happening, leaving more room to feel about it.

This is even more true of a soap opera, a form that is all about feelings. The currency of exchange is revelation of feelings and discussion of things that cause feelings. If characters are not talking about their feelings, characters in soaps are signifying their feelings in very recognizable and clichéd behaviors and visual cues.

Mad Men withholds most of the talk of feelings that we expect in the genre as well as the behaviors we are trained to read; *Six Feet Under* offers a more traditional treatment of words and signs providing access to the inner lives of the characters, even if the realities depicted are not precisely traditional television fare. With *Mad Men*, viewers are left with scenes that require attention to a less recognizable vocabulary of behavior to try to understand what the characters are feeling.

The result is that viewers watch and ponder why Don is so rude, or why Peggy made a horrible mistake even though she did what seems to the audience to be the right thing. Like watching a foreign film in a language you studied in high school, *Mad Men* requires attention in a way that few other television shows do.

Increasingly, longform television shows have used a strategy of limited access to the inner life of the lead characters, especially in mysteries like AMC's *Rubicon* and *The Killing*. Interestingly, both came in for much criticism for their slow-developing stories and for not doing television the way it should be done.

The minimalism of this storytelling, more obviously seen in *Mad Men*, seems to be an emerging literary style of television storytelling, a limit to the omniscient access we expect in a television show, engaging the viewer by whispering rather than shouting.

While other factors may be involved in the dissatisfied audience reaction to *The Killing* and *Rubicon*, the unhappiness could be understood as a consequence of withholding the inner lives of the lead characters, forcing the viewer to try to figure out what is happening, rather than having it explained. As these series are released as DVD sets, I expect a more positive reassessment as focused, extended viewing will turn this broadcast "bug" into a novelistic "feature" as more immersion offers more information.

The Good Wife: Woman's Soap in a Man's World

The most interesting use of the withheld inner life of a lead character may be found in the first season and a half of the critically praised CBS show *The Good Wife* (2009–). On one hand, the series is a potential rival to the shows discussed at the end of Chapter 4 as examples of the successful balance of the long arc and the single episode. On the other, it may be the nontragic balance to *Breaking Bad*'s overt depiction of a character transformed as she comes to terms with the consequences of decisions that she has made.

That is to say, standard novelistic storytelling.

The Good Wife follows the life of a woman coming back into her own after being publicly humiliated as the wife of a Chicago district attorney who is convicted of corruption and revealed to have had an active, adulterous sex life with prostitutes. The wife, Alicia Florrick (Julianne Margolies), is forced to return to practice law as a junior associate at the firm of her law school boyfriend, raising her children and getting on with her life as her husband, over the first two seasons, successfully appeals his conviction, moves back in with her and the children, and successfully runs for reelection. On election night, she discovers that he slept with her best friend and she leaves him (actually moves him into an apartment for which she has paid three months rent) and ends the second season going to a hotel room with her old boyfriend.

Each episode tells a story that has a clear closure, almost always a legal case, but often there is a personal triumph or closure as well. At the same time, the intertwined long arc stories are advanced.

What is striking about *The Good Wife* is that it spends its first season with Alicia Florrick keeping her own counsel, and at the edges of other people's stories. For the first season she isn't even the center of the action. She is more a witness to events in her husband's defense and campaign, a witness and supporting player to different lawyers in the firm as they carry out big cases each week, and a protective observer of the reactions of her children to the chaos. For many viewers, her best friend, the firm's investigator, becomes the most interesting character as she navigates competing political, commercial, legal, sexual, and mortal complications while remaining a woman of mystery.

As the series proceeds, Florrick becomes steadily more central to the stories and more assertive, finding the key piece of evidence, having the right technique to soothe a client or guide a witness but, for the first season and much of the second, Florrick is constructed as a supporting player to everyone, heightened by episodes focused on the firm's troubles, individual crises among the partners, the dating and religious ideas of her children, and political maneuvering by her husband's advisers.

Most important, the series does not reveal her thinking or her feeling. Where the investigator is mysterious, she constantly provides information, taunts, and deflections. For most of the first season, Florrick reveals as little about herself as possible. She listens to other people talk and reacts.

At the plot level, this is the depiction of a woman burned by having her private life plastered through the tabloids who holds on to what little privacy is left her. At the formal level, it forces viewers to pay more attention to her to see the clues to her reactions and plans to what she witnesses. Situations and behaviors are more familiar on *The Good Wife* than *Mad Men*. Over time, it is clear how she feels about most people and events but there remain hidden areas and actions that are unclear. Florrick lets her husband sleep in the bedroom, but what that means is unclear to viewers as well as her husband who thinks he understands, but clearly doesn't. When she later rents him an apartment and moves him into it, delivering the key as a fait accompli, her anger is clear but her methodical response suggests something more. Whether that "more" is a sense of spousal duty, a desire to not burn all bridges, a belief in him as a politician, if not a husband, or a desire to protect her children, or something else, remains unclear.

By the end of the second season, her apparent passivity was quickly disappearing as Alicia Florrick moved closer to the center of the stories being told, with her active decisions that saved her husband's campaign and turned losses to victories in cases for the firm. She also threw out her husband, banned her mother-in-law, and went to a hotel room with her boss and former boyfriend.

At the same time, Florrick increasingly expressed her feelings. While the character remains guarded, the betrayal by her best friend and her husband make her more emotionally expressive and push her toward taking more control. In an off-hand comment to an Irish advocate in the third season about her dislike of change, Alicia reveals more about her emotional life than she does in the entire first two seasons.

The format allows for the same situations in each episode, a mix of family, friends, politics, and a legal case. But the goal of *The Good Wife* is a progressive transformation of character. That is, change.

The apparent long arc of the series has Florrick moving on to a full independent life as a single mother and partner in a law firm, probably running for political office, becoming more active and more expressive of her inner life. The series is intended to be a story of continual transformation, a dangerous idea in network television, where characters are never supposed to change.

So far *The Good Wife* is succeeding. *The Good Wife* and *Breaking Bad* suggest that in network and basic cable there is an audience—probably two quite different audiences—for a long story of a character going through a transformation, the opposite of traditional television narrative, but one that rewards immersive viewing on DVD.

Chapter Seven

Comedy on Television

In Which We Consider the Rules of Sitcom, the Soapiness of Syndicom, an Origin of Dramedy, and the Rise of the Attitudinal Comedy

Jerry: So everybody I know is a character on the show . . .
George: Right!
Jerry: And it's about nothing.
George: Absolutely nothing.

—Seinfeld[1]

Situation comedies have resisted the long arc better than dramatic shows did.

Or perhaps that should be put differently: Sitcoms have had even more pressure not to change, because they are often consumed as freestanding episodes, but also because sitcoms have been the basis of syndication, and are therefore unforgiving of long arcs and dramatic change.

These functions of the classic situation comedy contain a dramatic limitation; the individual episode must restore the balance back to the state at which the episode began. David Marc makes the case of the unchangeable nature of the sitcom protagonist in the context of *I Love Lucy* (1951–1957), which I will grossly oversimplify as the observation that every episode begins with Lucy starting a new venture that begins positively but then hits an obstacle that sends her crashing into a far worse position than she began, and the rest of the episode is spent watching Lucy trying to claw her way back to where she started.[2] This highly successful model of sitcom structure continues to dominate the situational comedy to this day.

Perhaps not surprisingly, the same limited narrative form has also been central to the comic book writing throughout its American development. The similarity allows us to consider the challenges of writing the form from a slightly different angle, which is provided by the most prolific comic book writer of all time, the late lamented Paul S. Newman.

Newman wrote, among others, *I Love Lucy* comics, *Man from U.N.C.L.E.* comics, and *Lone Ranger* comics and the *Lone Ranger* comic strip. Also

Darkwing Duck, Fat Albert, Roy Rogers, Tom and Jerry, I Spy, The Twilight Zone, and innumerable other licensed properties, including the official comic for *Yellow Submarine.*

Newman pointed out that writing licensed characters takes a special talent that is not especially prized in our culture. Traditional dramatic narrative has a dramatic arc based on the transformation of the central character. This is the characteristic often associated with longform—and "quality"—television. Writers of licensed characters—for comics or for most situation comedies—must create compelling narratives without changing those characters or their circumstances.[3]

Newman noted that if Lucy suddenly became aware of her situation, the series would be over. The challenge of writing licensed properties are the same of writing most American television—creating plots that enhanced the reader's enjoyment of an unchanging character, the entertainment arising from continually reinforcing, not developing, the characteristics of the character.

The narrative structure of traditional American television is revealed most clearly in the sitcom, but it's been the same for most dramatic television as well. Each episode presents the progress from a starting equilibrium, a situation requiring reaction, a complication that results from the reaction, and resolution as a return to the original state.

Sitcoms require progression without change. Without progress. The permanent middle, narratively speaking. Or the permanent present, with neither history nor future.

Seinfeld (1990–1998) famously embraced this quality in the show's supposedly radical working motto "no hugging, no learning"—officially, traditional sitcoms are full of valuable lessons learned, but for all the supposed learning, there is no change. At best, the *audience* is supposed to learn from the sitcoms, but the characters within the sitcom don't learn. Think of your favorite characters and how part of what makes them your favorite is that they repeat the same mistakes over and over again.

As *The Sopranos* (1999–2007) demonstrated, American television traditionally has been about no change. The sitcom only exaggerates that, as comedy is immediate, of the present. If the viewer needs to remember that a character quit smoking last season to get the joke, by the time they remember it, the moment will have passed.

The actual joke of *Seinfeld* is that the characters never change but don't pretend to. The show was about nothing—as most sitcoms are—but didn't hide behind a justification of the series that implied a learned lesson.

David Marc interprets this through a famous tight structural reading of *I Love Lucy* to show that the true lesson for viewers is not to challenge the status quo because any change you make fails, so it's better to never try. Except that failure makes for more entertaining television, so the characters must try and fail, try and fail, over and over again.

While it doesn't apply to every sitcom, it can be disturbing to discover just how many shows a half century later continue to reinforce that depressing life lesson.

Over several seasons, some change is inevitable, but it is incremental or drastic but inconsequential. The show might move to Rio but the characters and situations remain the same. Rachel in the first episode of *Friends* (1994–2004), escaping from her wedding and running into a coffee shop where she meets the cast, is quite different from the Rachel at the end, and yet she is still essentially the same in terms of the function she serves/what she stands for in the series. One season Rachel pursued Ross instead of Ross pursuing Rachel. But you can watch any episode and enjoy it because the point is the two of them belong together but there is always an obstacle to them getting together. Even when they are together, *they* become the obstacle to them being together.

George gets a job with the Yankees, Elaine loses a job with Peterman, but, as Jerry realizes about that time in a very special episode, everything in his life (and in the show) evens out. The balance and the roles remain the same. You can watch one episode from the first season and one from the seventh and get everything, because only the situations have changed.

It is a demonstration of the writing skill involved in *Friends* that Ross and Rachel go through so many variations on the frustrated lovers theme without actually changing the characters or the dynamics of the series. Even when one or the other is in a deep committed relationship, the audience knows it is doomed because in the pilot they saw that Ross had a crush on Rachel, so they know that the two will end up together in the end.

This point becomes embarrassingly—or amusingly—obvious with *The Simpsons* (1989–), assuming a viewer stops to think about it. The characters never age and the storylines have required the kids to attend the same grades repeatedly over the decades. As someone who has aged from a Bart-aged identification to a Homer-aged identification during the run of the series, I can testify that the characters didn't change, the viewer did.

Spotlighting this reality and the importance of syndicated reruns to the experience of watching the series, *The Simpsons* ended the second episode of Season Nine—"The Principal and the Pauper"—an episode whose main storyline was a parody of *The Return of Martin Guerre* (1982)—with Principal Skinner revealed to be Armin Tamzarian, an impostor who took Skinner's place upon returning from the Vietnam War. At the end of the episode, the town collectively decides to forget everything, never talk of it again, and to continue calling Tamzarian "Skinner."

Like the ending of *The Sopranos*, this is a wink at the informed viewers, acknowledging that no significant change can take place on a television show. If, after that episode, Principal Skinner was referred to as Armin Tamzarian, that would disturb syndicated viewing because syndicated reruns are shown out of order. Some days the character would be called Skinner, some days Tamzarian. So everything must return to normal by the end of the episode.

According to *Simpsons* websites and scholars, this is one of the most reviled episodes in the series' quarter-century history and one the creators regret making. It would seem that *The Simpsons* viewers can tolerate any manner of mockery of the series, killing off most of Springfield in annual "Treehouse of Horror" Halloween specials, thrusting Homer into the three-dimensional real world and even a "Behind the Scenes" documentary episode about the "real life" of the characters when they aren't making the series, but fans can't stand to be reminded of how episodic comedy cannot change if sitcoms are to work.

Tho Sitcom as Likeability Machine

The other lesson to take from *Seinfeld* is that simply the form of the sitcom determines our response to their characters.

Jerry Seinfeld and Larry David made the characters deliberately unpleasant people—narcissistic, judgmental with a lack of self-awareness, and no concern for anything that does not offer a direct and immediate benefit to them. Objectively, these are not people that most viewers—or any secondary characters on the show—would want to spend much time with.

But these unpleasant characters were on a situation comedy and what they did made viewers laugh. Over time, they became not just likeable or even "lovable," Jerry and Elaine, George and Kramer became "beloved." They were no less obnoxious, they had just been obnoxious on a situation comedy.

The sitcom has been the location of television's ideal families since it moved from radio. *The Goldbergs* (1949–1956) and *I Remember Mama* (1949–1957) reminded viewers that Americans were families of immigrants, and through much of the 1950s showed America defined as a melting pot of working-class ethnic diversity in the modern city. By the end of the 1950s, the urbanity of *Amos & Andy* (1951–1953), *The Honeymooners* (1953–1956), and even *I Love Lucy* had been replaced by the new American ideal of suburban families where *Father Knows Best* (1954–1960) and *Ozzie and Harriet* (1952–1966) remind viewers that everything is WASP-ish, middle class, and fine. By then, the lessons that Americans had absorbed were that the sitcom showed you what was normal and if your family wasn't like sitcom families, there was something wrong with you.

Ever since that time, even as the family has been replaced by the office family or the "family of choice," the dominant characteristic of sitcoms has been the likeability of the characters. Dabney Coleman may have been unlikable in *Buffalo Bill* (1983–1984) but that was only because the show was cancelled before viewers got to know him. (After all, his assistant, played by Geena Davis, liked him and she can't be wrong.) American sitcoms make characters likeable.

Consider Archie Bunker, the extreme example for American television in the 1970s. While Bunker on *All in the Family* (1968–1979) was a watered-down version of Alf Garnett, the opinionated bigot from the British sitcom *Till Death Us Do Part* (1965–1975), Bunker's offensive opinions and verbal abuse were a shock to American audiences when the show premiered in 1968.

Within a couple years, Archie Bunker was one of the most popular characters in American culture, with T-shirts, books, and bumper stickers (the apps of their day). Most important, audiences had shifted to laughing with him, not at him. Partly, this was a cause-and-effect interaction, with Norman Lear softening Archie to meet audience expectations, making him more lovable as the series went along. By the time *All in the Family* was over, and Carroll O'Connor continued the character on *Archie Bunker's Place* (1979–1983), Bunker was a softie, totally presentable in any American home.

The principle might be illustrated by thinking of a sitcom as a rock tumbler. Sitcoms start out with sharp-edged, abrasive characters that don't fit together. The combinations of characters are often described as making "sparks fly." The ideal, over time, is that all the characters in a sitcom are smoothed out by rubbing together and till they're all smooth, with no rough edges, and they emerge polished—which is to say "lovable."

Consider Louie DePalma, the "cockroach" who was the dispatcher for the cabbies on *Taxi* (1978–1983) and was eventually revealed to be a man hiding his vulnerability because of a life of great pain, who loved, and was loved by the end of the series. Hot Lips Houlihan, the hypocritical killjoy ruins the lives of the good doctors on *M*A*S*H* (1972–1983) in the first seasons, is their best buddy in the end. Or Lou Grant, the alcoholic scary boss who terrified Mary Tyler Moore (playing Mary Richards), who viewers wanted Mary to marry in the final season of the show.

In part, this is the result of the need for acceptable transformations in sitcom writing, but it is also expected by viewers and enhanced by the way that viewers watch sitcoms, especially in syndication.

Probably the main reason that the much-promoted *Seinfeld* finale failed to satisfy its audience—besides a promotional buildup that nothing could have lived up to—was that the show went out of its way to remind us of how unpleasant these "beloved" characters actually were, going so far as to put them on trial for being horrible people who watched a man get mugged and just laughed at him. The trial presented the opportunity to bring back characters from the entire run of the series and their testimony was just like the joke in *The Simpsons'* Armin Tamzarian episode, reminding viewers of what they must ignore to enjoy sitcoms. The witnesses testified that America's beloved sweethearts were horrible people.

After telling America it was in love with despicable characters, *Seinfeld* went one step further and ended with George complaining about the button on a shirt being in the wrong place, the very first lines from the pilot of *Seinfeld*. Not only did the characters remain unchanged, and even as they were sent off to prison, their dialogue didn't change.

Viewers enjoyed the reminders of past episodes but wanted to laugh with their television friends at the same time the script was telling them that they were wrong to love Jerry and Elaine, George and Kramer. It's worth noting that when they did the Seinfeld final episode do-over on Season Seven of *Curb Your Enthusiasm* (2000–), the script went out of its way to show that Larry David was a horrible person, but the *Seinfeld* actors and characters were as funny as viewers wanted to remember them.

Dead Men Laughing: The Laugh Track

Network sitcoms have laugh tracks. What most of the longform sitcoms have in common is the lack of a laugh track, thereby allowing the viewer to decide when or if to laugh as they watch the suffering of the central character. This simple formal shift is more telling than it appears.

The laugh track reduces anxiety, tells viewers what they should be laughing at, and assures them of their normality in laughing. The basic principle behind the laugh track is that people will laugh more if they hear others laughing, but they will also feel better because they feel the comfort of being part of a group. This is part of what Sigmund Freud referred to as the "oceanic effect," in *Civilization and Its Discontents*.[4]

Traditional television was based on making viewers feel a part of something. The iconic image of viewing was the family gathered together to watch television in the living room. We still refer to the first evening hour of national programming as the family hour, although the meaning of it has largely disappeared.

Sitcoms with laugh tracks are the audio version of *Mr. Rogers' Neighborhood* (1968–2001), creating the illusion that viewers are part of a community. In different ways, *American Idol* (2002–) and talk shows hosts such as Craig Ferguson and Jimmy Fallon have continued this dynamic. They offer membership by voting, or by reading viewer email, or by involving viewer surrogates from the audience into the family activities of the show.

The laugh track can also be seen as an effort to hold off the recognition of a shift in the meaning of television viewing. If television in the 1950s was the family huddled together around the television hearth, in the 21st century, it is the individual hunched over the computer screen.

Television has long had an element of isolated viewing to it, but it was disguised by the water cooler moment phenomenon, the idea that people may watch alone but everyone watched the same shows so that they could discuss them at the water cooler at work the next day. TiVo and Internet viewing, along with the availability of DVD sets, have made it increasingly antisocial to discuss a show because those who will watch it later don't want "SPOILERS." Water cooler moments been replaced by water bottles and spoiler alerts.

The loss of a laugh track is an acknowledgment of the transition to an isolated, rather than social, experience of television comedy. In conceptual terms, the shift from family to isolated viewer places television closer to the world of the novel—where an individual escapes into a fantasy as they read, as opposed to a film or a theatrical experience, where the shared reaction is integral to the experience.

Traditional sitcoms are filmed before a live studio audience with three cameras and, after the fact, "sweetening." Sweetening is the term for mixing in prerecorded laughs, titters, and guffaws to the sounds the audience made during the actual performance. This allows the producers of the sitcom to get the laughs they wanted but didn't get from the audience and teach the viewing audience that what they are watching is much funnier than it actually was. A good job of sweetening a laugh track is like the soundtrack to a movie, telling the audience how it is supposed to feel with being noticed.

*M*A*S*H*, was a significant shift in American sitcoms because the show was being shot without a studio audience so there was no authentic laughter, it all had to be completely added in postproduction. The network insisted on a laugh track for the first few years, so that the audience would know it was funny but, as the show became a commercial and critical success, the creators were allowed to remove the laugh track and it continued to be a hit.

Sports Night (1998–2000) successfully demonstrated the advantages of working without a laugh track in 1998. The show was able to create a density of fast-talking banter without having to pause for laughs, resulting in a few more minutes of dialogue per episode while making viewers feel smarter for laughing on their own. The availability of VCRs compensated for the fast dialogue and lack of cues as to where the jokes were, because viewers could back up and watch again if they missed something.

Comedies without laugh tracks lean toward the longform story, and the values often found in other longform stories on television. They tend to speak to isolation, even if community is seen as a good thing, and the comedy tends toward comedy of attitude, not situation comedy.

Without a laugh track, we are on our own.

The Soapiness of Syndicom

The sitcom is a form even more resistant to a longform narrative arc than television drama, but it arrived there, nonetheless, by a few different routes. To paraphrase Shakespeare, some are created with a long arc, some achieve the long arc, but others have the long arc thrust upon them.

Syndication changes a sitcom. The way the show is viewed changes what it means. Syndication changed the frequency and density in which viewers consumed television episodes, forcing it into a longform narrative without the choice of its creators.

Originally, the hours before prime-time broadcasting were offered to local affiliates to increase local programming in the interests of using the airwaves for the good of local communities. Affiliates quickly realized that the hours could be used more effectively to make profits by running inexpensive game shows and reruns of sitcoms. In recent years, the return to local news as a low-cost profit center for television stations has shifted the balance to some extent, but the expanding rerun heaven that is cable offers viewers many opportunities to watch reruns of shows like *Friends, Seinfeld, How I Met Your Mother* (2005–), and *Two and a Half Men* (2003–) in pre-dinner syndication.

Syndication is the economic goal of situation comedies, which explains the non-numerological obsession with getting to 100 episodes of a series, the traditional threshold for being viable in syndication. *Big Bang Theory* (2007–) set a new record when it sold its syndication rights to TBS and Fox in 2011 for more than $2 million per episode. It is estimated that just the production company—leaving out all other creative participants including writers, directors, actors, and so on—has so far made $4 billon from syndication payments for *Seinfeld.*

The economics of syndication created distribution patterns that turn sitcoms into longform stories.

If a viewer watches an episode of a sitcom once a week, the viewer gets their laughs and a sense of the relationships between the family units of the show. If a viewer watches the series five days a week in reruns, the same series start to change. Even more so if it is two—or four—episodes every day. And even more if the viewer is rewatching an episode that they've already seen.

The sitcom becomes a soap opera.

In syndication, *Roseanne* (1988–1997) becomes the working-class soap opera that Americans never got from the networks. The strains on Roseanne and Dan's marriage become more obvious, the fears about money, and about where their kids will end up, the difficulty in living with continued disappointment and continually disappointing those around you become developed themes in longform, where they are just situational setups for the jokes in episodic viewing.

Friends becomes more about the inability of the people to find fulfilling jobs or relationships—Chandler's recognition that a paycheck isn't enough; Monica's gradual recognition that she may not become the chef she assumed she'd be; Ross's difficulties parenting a child who lives with his ex-wife and her lesbian lover and his self-destructive fixation on Rachel become more pathetic; Rachel's lousy taste in men; and Joey's inability to act, interrupted by unjustified career success.

This is the most neglected way in which dramedy came into being, not as an intentional creation, but as the inadvertent side effect of rerun viewing. Binge viewing of sitcoms—especially of episodes already seen—makes the dramatic story in the narrative emerge from the comedy. When you've already heard the jokes, the relationships remain. All sitcoms can become soapy under these conditions, but the 1970s sitcoms that went into high rotation in syndication, like *M*A*S*H, WKRP in Cincinnati* (1978–1982), *Mary Tyler Moore* (1970–1977), *Taxi, Happy Days* (1974–1984), and later *Cheers* and *Friends*, where relationship-based emotional moments occasionally supplemented the laughs, made it all the easier for the experience of daily viewing of a sitcom to replicate the experience of daily viewing of a soap opera.

Film editing is based on the fact that humans are hardwired, when faced with several discreet moments, to make connections between the moments in search of meaning. A series of shots are automatically interpreted as related to each other and, possibly, as a progression. (This basic principle is known as the Kuleshov effect.) Watching daily episodes of a situation comedy, even if the series has no intended longform story, engages the same pattern recognition behavior. It becomes natural that viewers to connect similar moments in different episodes and construct meaning from—or for—them.

Hot Lips on *M*A*S*H* is an absurd character—a career military strawwoman for the cool doctors to work against. But in more intense rerun viewings, even in the first seasons, she becomes a woman trapped in a limited role, forced to find sexual comfort in a weasely married man who will never leave his wife. In later seasons, the writing of the show changed the depiction of Hot Lips (dropping the nickname for Major Houlihan or Margaret) to overtly address the difficulty of her position, but it was already obvious to syndicated rerun viewers. Houlihan's first season behavior becomes far more sympathetic and complicated when an episode that comically condemns her coupling with Major Burns is immediately followed by another episode that applauds Hawkeye's jumping into bed with a new nurse each night.

This happens even if the series rejects longform or even continuity. *Taxi*, one of the most highly praised of the 1980s sitcoms, created each episode to pay off on a tight three-act structure with no concern about story continuity between episodes. Tony the boxer spent a couple episodes adopting a foster child only to never be mentioned again. Louie dated Zena (Rhea Perlman), they got very serious, then she was gone, Louie was dating others, then suddenly Zena was back, only to be gone again. Such continuity flaws don't matter in weekly episodic television, but watched back to back the show weakens.

Even so, *Taxi* still ends up leaving syndicated rerun viewers with many soapy questions to ponder—whether Alex and Elaine will ever get together, whether Latka can assimilate without serious psychological damage, whether Tony will find a way to turn his failure at boxing into a meaningful career, whether Bobby can—or should—succeed as an actor, and whether all of the characters can find love. In syndication, the close juxtaposition of Reverend Jim being treated as an idiot and Jim's caring but unsuccessful relationship with his father add an implicit explanation for his attraction to the abusive father-figure in the dispatcher, Louie. And those relationships make a few widely scattered jokes by Jim expressing his anger suddenly far more meaningful insights into his character, whether that was the intention of the creators or not.

The practice of American sitcoms making all characters likable over time trains viewers to care about the characters of sitcoms as more than joke delivery systems. More immersive viewing narratizes that response.

The Sitcom Soap Opera

The more formal way that dramedy was used to intentionally create long-form television comedy was by doing what Steven Bochco did with *Hill Street Blues* (1981–1987). *Soap* (1977–1981) is the sitcom equivalent of *Hill Street Blues*, consciously incorporating the soap opera format into the situation comedy. Except *Soap* did not just absorb characteristics of the soap opera, it explicitly parodied the form. Shot on the same sorts of sets as the daytime dramas, the show took soap opera types and their plotlines to the absurd extreme. The same sorts of characters (some played by former soap opera actors), themes, and acting were now used to set up jokes, but the audience also followed the series with an increasing investment in the stories of the families of two sisters—one rich, one middle class. The series rejected freestanding episodes and told one continuing story of the Tates and Campbells that Susan Harris, its creator, intended to wrap up in the next season when it was canceled. Viewers could watch the soap ironically, but they were still getting the daytime drama storylines.

Arrested Development (2003–2006) owes much of its structure to *Soap* (including the device of the cheerful introductory voice-over narrator), but *Arrested Development* uses the documentary film style, making a formal claim of realism for the ridiculous characters by shooting them in real locations rather than the soap opera studio. The result is that the comedy shifts from *Soap*'s parody of the absurdity of soap operas to a blend of soap opera narrative and fake documentary form to tell stories about absurd characters to an audience that may not know the undiluted soap opera references anymore.

Mary Hartman, Mary Hartman (1976–1977), created by Norman Lear, was the first, and more demanding, version of what *Soap* did. Rather than simply doing an open parody of a soap opera, the show was an actual daily soap opera in sitcom form. Plotlines were tweaked into absurdity but without overt markers to tell the audience when—or whether—to laugh. The show ran nightly in the same syndication time slot that viewers tuned into for their reruns, often late at night because of its controversial content.

The show told the longform, serial story of Mary Hartman (played by Louise Lasser) and the people of Fernwood, Ohio. Unprepared viewers would notice there was something off with this soap opera. The problems were a bit too absurd, the reactions a bit too existential, and the manner in which people died was blackly comic (one character drowns in a large bowl of Mary's chicken soup as she rabbits on about its health benefits). The series was as much a deadpan expose of the absurdity of soap operas as it was a show you could tune in to get laughs, and it shared a similar unsettling vibe with *Twin Peaks*, another simulacrum of a soap opera in addition to its other plots.

To confuse the limits of comedy even further, *Mary Hartman, Mary Hartman* dispensed with soap opera euphemisms and directly addressed by name spousal abuse, drug abuse, incest, sexual deviance, and other problems only hinted at on daytime television. The result was on one hand more realism, and on the other more absurdity.

Negotiating the balance was the seriously depressed Mary Hartman who alternated between deadpan bleakness—explaining her houseplant is dying because it's taking on the personality of its owner—while also expressing the concerns of the television housewife about unsightly yellow buildup on her floors.

Mary Hartman, Mary Hartman became a fascinating experiment with how far comedy could be stretched on American television, clearing ground for *Twin Peaks*' (1990–1991) ironic sincerity and spun off the more accessible syndicated fake talk show, *Fernwood 2-Night* (1977), hosted by Martin Mull and Fred Willard.. The type of comedy these shows mined ends up working best on film, with the faux documentary (or mockumentary) movement associated with *This Is . . . Spinal Tap* (1984) and the films of Christopher Guest, especially *Best in Show* (2000) and *A Mighty Wind* (2003), where the culture that is the subject of the humor, dog shows and folk music reunion concert documentaries, are so painstakingly recreated that large sections seem more drama than comedy, with the laughs that occur buried deep within the recognition of personalities, not in overt jokes.

The most influential absorption of the soap opera into sitcom was a critical darling that was never a commercial hit, *The Days and Nights of Molly Dodd*. Canceled by NBC after two abbreviated seasons in 1987 and 1988, the show was revived with three more short seasons on the Lifetime channel following the success of the series' reruns on the channel.

Shot with a single camera, with multi-episode arcs, no laugh track, and plenty of minor tragedy, the series' influence continues to be seen in long-form comedy, in establishing the second intentional type of dramedy, and in providing the template for a new type of television comedy: the attitudinal comedy.

The Attitudinal Comedy

The Days and Nights of Molly Dodd (1987–1991) started as a midseason replacement series on NBC in 1987. A critical hit, the show could not hang on to an audience. *Molly Dodd* tells the story of a divorced New Yorker in her thirties trying to figure out what to do with her life, about as far from Lucy Ricardo (another New York woman trying to figure out what to do with her life) as one could imagine, she struggles with an ex-husband, parents, dating, and careers, including a breakdown leading to psychotherapy, death, and unwed pregnancy. Her life is filled with indecisions, bad decisions, and disappointment.

When watched in the density of daily reruns, the soap opera qualities become overwhelming, but they are already obvious in weekly viewing. The story of Molly's struggles renders the laughs empathetic, a way of processing the sadness of the disappointments of existence.

Without a laugh track to guide viewers, the show functions as a new type of soap opera as much as a comedy. Instead of the traditional soap's closed community full of related, if not inbred, people driven by passions to love and betray, here viewers follow a single woman mainly unconnected and, increasingly, unable to escape her parents, even after their death, as she pursues connection and meaning.

What's inescapable for Molly Dodd is present in most of the other syndicated sitcoms of the era, but there it's deflected and disguised. To understand how *The Days and Nights of Molly Dodd* is a change in attitude, rather than content, it is worth considering how much of the subject matter on the series was also covered in a far more traditional way on the show *Seinfeld* or *The New Adventures of Old Christine* (2006–2010). In both series, Julia Louis-Dreyfus plays women experiencing wackier versions of the problems that plague Molly, with laugh track–enhanced situations.

The Days and Nights of Molly Dodd's difference is apparent as Molly's mother's judgmental narration—not Molly's own voice—opens each episode, reinforcing Molly's inability to control even her own story, let alone her life. This variation on narration only adds to viewers' identification with Molly's point of view because, even as they reject the mother's judgment, they start each show with that criticism ringing in their ears, bonding them more tightly to Molly. The comedy is more natural, droll comments rather than jokes. The laughs are private, not shared with a laugh track. Viewers get to decide where they think the laughs belong.

While it was called a situation comedy, the comedy of *The Days and Nights of Molly Dodd* is not especially situational. Instead, the comedy is attitudinal. Traditionally sitcoms are based on a wacky event or misunderstanding that generates laughs, funny situation. Molly Dodd has far more depressing situations than funny ones.

The humor of this series is found in Molly Dodd's attitude toward her inescapably depressing life. Essentially the narrative logic is soap opera, but it's not just the story of a noble character's continuing perseverance in the face of disappointment, it is the story of her droll comic attitude toward her suffering, her ironically detached resignation that creates the comedy. From the beginning, sitcoms have included words of authoritative wisdom that *tell* viewers how to live; Molly Dodd *demonstrates* how to live.

Attitudinal comedies present "normal" events of life enhanced through the lens of melodramatic suffering. The humor is found in the acceptance of reality shaped by the lead character's attitude toward what is happening. One of the progenitors of attitudinal comedy would seem to be George Burns who overtly narrated *The George Burns and Gracie Allen Show* (1950–1958) where viewers would observe the wackiness of Gracie Allen and then hear George Burns ask, essentially, "Would you look at this? Can you believe this is happening?" But Burns commented on weirdness, not despair. *The Days and Nights of Molly Dodd* takes the point-of-view of the titular character, asking us to see the absurd humor in her life of powerlessness and disappointments.

Not surprisingly, this attitudinal form of comedy has not prevailed on network television. However, it has become the dominant form of comedy on cable television. Not simply the existentialist comedy of attitude, but also the comedy of female-centered (and occasionally "feminized" male-centered) suffering.

Chapter Eight

Longform Comedy on Television

In Which We Consider British-Style Sitcoms, then the Longform Attitudinal Comedy and Gender, Documentary and Reality Comedies

He's got to care, or it isn't interesting. He's got to fail or it isn't funny.

—Ricky Gervais[1]

A Short British Longform Tangent

British sitcoms have long been built around the assumption of a single season (called a series in the UK). A writer or writers would pitch a series with the assumption of a single, probably six- or eight-episode, run. Without the goal of producing a series that will run until the end of time and make everyone involved a millionaire, this approach encouraged experimentation with the form, and created a natural ground for individual writers to tell novelistic series with transformation and endings.

If a series succeeded, they might be asked to follow it up. While hit shows have led to long runs on British television, for every long-running soap-ish sitcom such as the original British *Shameless* (still running in series of increasingly larger numbers of episodes—the last being 22 episodes—after eight years on the air), there are also short-running longform successes such as *The Hitchhiker's Guide to the Galaxy* (1981), *Blackadder* (1982–1989), *The Biederbecke Trilogy* (1985, 1987, 1988), *The Fall and Rise of Reginald Perrin* (1976–1979), *The League of Gentlemen* (1999–2002), and *The Office* (2001–2003). In the realm of episode-centered sitcoms, a three-series *Black Books* (2000–2004) or a two-series *Spaced* (1999–2001) or *Fawlty Towers* (1975 and 1979) are still considered great successes.

The difference allowed by the short series is best illustrated by the fact that they are often written by an individual or pair of writers, as opposed to a room full of writers. *The Fall and Rise of Reginald Perrin* was a broad situation comedy adapted by author David Nobbs from his darker 1975 novel, *The Death of Reginald Perrin*. The series tells the story of a middle-aged bureaucrat at a dessert company who begins to rebel against the banality of his job and his life. Eventually he fakes his own suicide to start his life over again in various absurd disguises. Before long, he has fallen back into his old life under the name of Martin Wellbourne, with no one commenting on the fact that he is obviously Reginald. As Martin, he marries his own widow and ends the first series where he began.

Nobbs wrote two more series and two more novels of *Reginald Perrin* (1977, 1978–1979). In the second series, Martin decides to waste his life by starting Grot Shops, shops that sell only useless "grot" (slang for crap), which makes him an unwitting multimillionaire. Everything he does to sabotage his company—including hiring his relatives and fellow employees from Series One's Sunshine Desserts and making increasingly self-destructive business choices—only ends up increasing his success. The second series ends with Reggie and his wife going to fake their suicides. The third series has Reggie opening a commune for unhappy middle-aged people and when that collapses, he goes to work for his ex-military militia-oriented brother-in-law. Faked suicides are involved in the series finale. The *Reginald Perrin* series each lasted seven episodes, about three and a half hours, less than a third the length of an average American sitcom of that period.

The Fall and Rise of Reginald Perrin is an overtly novelistic series, written by the author of 17 novels who also wrote on a dozen television series. While there is pleasure in any single episode, the full narrative effect only arises through escalation, as the viewer watches Reggie's increasing desperation and rising alienation from middle-class English culture, as the insanity of situations naturally grows from one absurdity to another, the consequences and repetitions across episodes and, through it all, the struggle of his reasonable wife to allow him to have his little rebellions.

In 1978, *The Hitchhiker's Guide to the Galaxy* premiered on BBC4 Radio, was re-created as a television series the next year and then turned into a series of novels. "H2G2," as Neil Gaiman renamed the series, is exemplary of British approaches to the sitcom because the series—a single serial story of a man, Arthur Dent, who is rescued from the destruction of Earth by his best friend, an alien named Ford Prefect, leading to a space adventure with the two-headed fugitive president of the galaxy, a woman Arthur once met at a party, and a suicidal robot—began as a radio series, became a television series, returned in other radio series (all of which were adapted into novels), was adapted into one of the first (text-based) video games, and was eventually made into a film.

While *The Hitchhiker's Guide to the Galaxy* did it more completely than most, in British television based on novelistic storytelling, there is a far greater likelihood that a longform series will cross at least one media platform.

In the 1980s, Alan Plater created his most extended body of work that is called *The Biederbecke Trilogy* in DVD set release. Plater adapted his Thin Man–inspired novel into *The Biederbecke Affair* for ITV (1985—six television hours). That was followed by *The Biederbecke Tapes* (1987—two extended episodes) and *The Biederbecke Connection* (1988—four television hours). Each series follows the low-key romance of a middle-aged Yorkshire woodworking teacher/jazz enthusiast and an English teacher/environmental activist who navigate spy genre plots made comic through mild, absurdities and the witty banter of the two lead characters. Plater's series might be best compared to Garrison Keillor writing novels about intrigue in Lake Woebegone.

Of all of his work, the 1995 short series *Oliver's Travels* may be the most successful as a poignant later-life romance between two witty people played by Alan Bates and Sinead Cusack as they solve the mystery of a missing crossword puzzlist on a road trip that takes them from Wales to the Orkney Islands. The basic model is a screwball comedy film, but the additional time—about four hours without commercials—and leisurely pace changes the madcap wit-assault of that genre into the gentle slide of two older romantics on the run from international corporate intrigue, as quirky and threatening events and people interrupt them.

The key distinction from American television in these series is that each series was written by a single author, who wrote and plotted television as if it were a novel with a beginning, middle, and ending advanced through episodic chapters. Looking outside of U.S. television reminds us of how unusual the American group writing approach is in the world of commercial narrative. Theater, film, novels, graphic novels, and operas are generally created by a single writer or a small collaborating team.

The British model of comedy writing influenced the creation of comedy series in Canada and Australia, with Canadian hour-long longform comedy series creating an interesting deviation from U.S. dramedies. *Slings & Arrows* (2003–2006), probably the best example of this Canadian approach, is known to most U.S. audiences as a seven-disk DVD novel, although it ran on Sundance Channel and then reran on Ovation. Like *The Fall and Rise of Reginald Perrin* and *The Biederbecke Trilogy*, each episode of *Slings & Arrow* must be watched in order to follow the story of the trilogy through its run of 18 television hours.

The series defies categorization. In some ways, it's an hour-long sitcom, but it's a comedy that turns steadily more serious over its three seasons, a transition that is rare in television—in U.S. television, only *Breaking Bad* (2008–) has followed a similar progression. Dramedy does not adequately describe the continuing tonal transformation of the show. Perhaps "situation drama with comedy" would be most precise.

Slings & Arrows takes place in a fictional version of the Shakespearean Stratford Festival in Ontario. Oliver Welles, the artistic director of the New Burbage Festival, gets run over by a truck hauling Canada's Best Ham and he is replaced by his former protégé Geoffrey Tennant, a once great young actor-turned-director who is considered insane.

Geoffrey is played by Paul Gross, best known for his work as the Mountie-out-of-water in Chicago in the series *Due South* (1994–1996). He returns to the New Burbage Festival as its artistic director, only to be haunted and harassed by Oliver's (Stephen Ouimette) ghost for the rest of the series. Tennant is also saddled with major budget problems and a duplicitous business director (Mark McKinney) who, with his American corporate girlfriend, wants to take over the festival, throw out the classics, and just do musicals.

The series was designed with a three-season structure. For each season, there is a centerpiece production of one Shakespeare play and the long-arc plot of that season is a variation on the themes of that play. Season One is *Hamlet*, Season Two is *Macbeth*, and Season Three is *King Lear*. And the struggles of youth, established adulthood, and old age resonate through each season, turning from light toward increasing darkness by the end of the third season.

The series starts essentially as an hour-long situation comedy but the jokes turn to laughs of recognition and resignation over the run as the series turns more serious. The first season is built around a fictionalized version of Keanu Reeves's actual appearance as Hamlet at the Stratford Festival in 2000. Young love, jealousies and failed love affairs, the pretentiousness of theater people, the evil of American corporate lovers of musicals, and the wacky misunderstandings as Geoffrey insists on publicly arguing with the ghost of Oliver provide the laughs. Very serious threats and drama occur from the start, but the villains of the first season are a Canadian buffoon and Canadian caricature of a U.S. femme fatale.

The best argument for considering the whole series as a comedy might reside in the obvious Shakespearean roots, and the traditional idea of comedy as "lovers united" at the end, rather than the importance of jokes. Characters are amusing, and one true clown, Darren Nichols (Don McKellar's avant-garde director/poseur), arrives each season, but the humor—like the arc of the entire series—is more of the sort one expects in a witty literary novel rather than a sitcom.

Over the seasons, the youngish artistic director becomes old in mind more than body, as characters adjust to their own limitations. The buffoonish avant-garde director becomes less ridiculous, the business manager becomes sympathetic and but turns out not to have lost his venal side, and attaining enduring love and making great art, even if it is unrecognized, become the only triumphs in the end.

Slings & Arrows reveals the superiority of the long-arc television story to film, the ability to develop story over 18 hours, to set up echoes between the central play and the season, but also echoes across the seasons. Events in the third season only have their full effect if you understand—and remember—what victories were won, and how, in Season One.

What sets the series apart from U.S. work, though, is the treatment of change and aging. The struggles of the protagonists over the series are the struggles to cease to need to be a star, to be a nurturer and guide for others, and to find a way to live contentedly in support. There may be something Canadian about this theme. Robertson Davies's novel *Fifth Business* comes to mind as an example of this Canadian lesson learned, but it the theme of transformation is uniquely suited to longform television, and especially the DVD novel.

The Attitudinal Comedy

While no U.S. comedies have attempted the longform situation structure of *Slings & Arrows*, or the closed forms of the British comedies discussed above, the descendants of *The Days and Nights of Molly Dodd* (1987–1991) have some of the same thematic developments that set those series apart. On cable television, the half-hour comedies have generally attempted to tell longform adult stories, with season-long and series-long stories of gradual transformation. As mentioned previously, this shift is accompanied by a shift to attitudinal rather than situational comedy, with resigned acknowledgment being the source of humor rather than explicit jokes.

Cathy Jamison (Laura Linney) on *The Big C* (2010–), Jackie Peyton (Edie Falco) on *Nurse Jackie* (2009–), Tara Gregson (Toni Collette) on *The United States of Tara* (2009–2011), and Amy Jellicoe (Laura Dern) on *Enlightened* (2011–) all follow the model of Molly Dodd as they generate humor for the audience by surviving in the face of despair. There are very few jokes in these shows, more often than not the situations are not funny, they exist somewhere on a continuum between absurd frustration and inescapable nightmare, with pressures from all sides and a lack of empathy becoming extreme selfishness from those around them.

These comedies tell the viewer, "You are on your own." The absence of a laugh track reinforces that. The result is a very existential sort of laughter, laughter in the face of death and powerlessness, the sort of laughter that Camus invokes in discussing the dilemma of modern man in his "The Myth of Sisyphus."[2] Camus's essential argument is that life, like Sisyphus's task of pushing a boulder up a mountain for eternity, is meaningless. However, like Sisyphus, if a person recognizes the meaninglessness of their existence but chooses to continue to do it anyway, they become existentially heroic.

Where the episode-centered sitcom is intended as an escape from life's concerns, these longform comedies offer the humor of recognition and resignation. The strange thing is that the attitudinal comedy's heroine is much like the long-suffering protagonist of classic soap operas, isolated in the midst of community by heightened sensitivity and struggle. The laughter of the attitudinal comedy is a coping mechanism turned to the entertainment, the solution to Sisyphus's dilemma.

Life is a meaningless veil of tears but, seeing that, we might as well have a laugh.

Not surprisingly, these shows skew older in their demographics. They are aimed at adults with life experience, adjusting to lowered expectations. Functionally, this form of dramedy serves a similar survival function to that of soap opera. The traditional soaps, and the grand melodramas of the 1930s, 1940s, and 1950s on the big screen, showed us people with lives worse than ours and taught viewers (especially women) how lucky they were to have it as good as they did. These attitudinal comedies offer a slightly different balm; they teach the viewers (again, more women than men) to survive by finding humor in despair.

Gendered Comedy

One shared characteristic of many of these attitudinal comedies is that they offer a female perspective with a sardonic humor of resignation to an unchangeable situation. From a cultural perspective, it's a disturbing reflection of a shift away from female empowerment comedies, but it's also part of a narrative function that goes back to the daytime dramas of women's suffering.

Sex and the City seems to straddle the idea of the attitudinal comedy. Carrie (Sarah Jessica Parker) narrates the show in her sex column, following the model of Molly Dodd's attitude toward her situation being the source of droll humor, rather than jokes as such. The message of Carrie's narration is that "I am above the life I live. I am superior to others because, even though I make the same mistakes as they do, I do it with awareness and an ironic perspective." This comes off as a more empowered version of the Sisyphusian triumph of finding the absurd humor in one's situation.

Ironically, the television shows that develop from this attitude, though not *Sex and the City*, are the ones most likely to actually allow characters to change. Even with multi-episode arcs, *Sex and the City* remained firmly grounded in the sitcom logic of unchangeability and the closure of the freestanding episode, which was key to its easy move into syndication.

The women of *Sex and the City* are torn between two impossible ideals: individual independence and romantic bliss—throughout the series the characters verbalize a recognition of that fact but continue to act as if the two are resolvable. Change is the enemy in this show and biological clocks are ticking in a way that drives them toward relationships that will only lock into place as the series' ends. Up to that ending, the women of the series remain essentially unchanged—people could answer the question "Are you a Carrie, a Samantha, a Charlotte or a Miranda?" because they were such specific, unchanging types, even if they changed partners, or sexual preferences, or careers, they remained one of a group of women defined by their meaning in juxtaposition to each other.

The Big C is a more married and suburban version of *Molly Dodd* with a woman patiently caring for the people in her life after she is diagnosed with cancer. (A very different—though equally gendered—response to impending mortality from the male patient in *Breaking Bad*.) *The Big C* is essentially optimistic. After a first season in which everyone in Cathy's life was incredibly insensitive to her feelings about her cancer and obsessed with how her dying would affect them, the second season softens their supportive narcissism, but continues to find Cathy suffering in isolation in an insensitive world. Cathy perseveres through her ability to see the absurd humor in it all.

Nurse Jackie asks viewers to identify with the suffering of a flawed, adulterous drug addict whose home life is always falling apart while she is the most competent force in the ER. Even though she's objectively a "bad" person, her behavior is justified by her being surrounded by people who act worse, especially toward her. This strategy is also used in *Enlightened*, where a (formerly) horrible person has achieved at least the accessories of enlightenment but returns to a life where no one appreciates her desire for change.

The United States of Tara alters the model a bit, as it presents the flawed Tara struggling to remain in control of the alternate personalities of her dissociative identity disorder (DID). Tara and her nuclear family comment drolly on the pain in her life, and the pain that her pain has caused them. Those outside the family unit are insensitive and create problems, but the husband, son, daughter, and Tara perform the role of the existential victim turned hero through attitudinal humor.

The protagonists of attitudinal comedy have a sense of resignation about them. Chandler and the other *Friends* were not resigned, even if they might live in an absurd world where nothing ever really changed. Neither were the casts of *Seinfeld*, *Cheers*, or any of the network sitcoms. Their wacky jokes reflect a kind of control over their situations. The jokes of attitudinal comedy are the sorts of comments the viewers would make in every day life.

These are adult comedies—not in the sense of having more sexuality or controversial themes (although they might)—they are "adult" because their attitude is a humor grown from adult experience with the disappointment of life, the need to compromise and adjust one's expectations. The economics of cable television, with a more limited definition of commercial success, makes it possible to tell these stories not aimed at children.

Most of the families on television comedies have been seen from the child's point of view—because they are the targeted demographic—but these attitudinal comedies tend to favor the adult's view of family. *United States of Tara* is a family comedy in the true sense of the word; it is about how family works and doesn't, but the key dramatic throughline is the concern of the parents for their children, the acceptance that the parents' lives are screwed up, and the concern that it might pass on to their children. The parents are balanced by children who have learned to cope with their parents' limitations but remain optimistic about their own futures (even if viewers may have more concerns). *Roseanne* was one of the few network series with this dynamic, although the parental limitations were presented as mainly economic.

Weeds (2005–) offers an intriguing variation on this form. It's a story about a suburban single mother who becomes a marijuana dealer to pay the bills and discovers that she is very good at it. Structurally, it is comedy because, to compare it to a similarly themed *Californication* (2007–), Nancy Botwin never appears to suffer the consequences of her actions.

Even though Nancy went to jail, it was between seasons, so viewers didn't experience it. She will never completely succeed, but she will always survive, and drag her quartet of supporting men along with her. There are broadly comic supporting characters but the closer they get to Nancy Botwin, the less absurd they become.

It is still the attitude toward events that creates the comedy. Each male she gets involved with is voluble in reactions, often comic. But Nancy Botwin's character is a surprisingly unrevealed character, even after seven seasons. Played by Mary Louise Parker, she is likable, attractive, and flirty but we rarely get any sense of her values or thinking. In a variation of the attitudinal comedy that appears more frequently in the male-centered shows, she is objectively unlikable—a femme fatale constructed as an innocent, manipulating the men in her life as she plays the victim—but performed charismatically in a way that leads audiences to forgive an absolutely narcissistic character.

Nancy Botwin would be a typical protagonist for an attitudinal comedy if she were male.

Where the Boys Are

There are male-centered attitudinal comedies, of course. *Californication*, David Duchovny's longform fantasy of being a talented artist who young oversexed women throw themselves at, while a loving (technically, ex-) wife and daughter forgive him all his libidinous and chemically enhanced excesses. *Entourage* (2004–2011) tells the longform story of boys who take advantage of the best of everything in Hollywood as the entourage and manager of a mega movie star. *Hung* (2009–), the longform story of a male prostitute/part-time high school basketball coach, Ray Drecker (Thomas Jane), and his female poet/pimp, Tanya Skagle (Jane Adams), is a fantasy of male empowerment through prostitution in the middle of a bad economy.

The shared theme of these series is that all men are boys, all women are adults. This is a set of values not that different from the female-centered attitudinal comedies. The implied long arc of all of the series is the boys becoming men, but the shows focus on the pleasures of extended boyhood.

Entourage (2004–2011) presents the everyman reaction to the absurdity of real Hollywood, slightly tweaked, with the straight man "E" (Kevin Connolly), manager to buddy Vincent Chase (Adrian Grenier) as our attitudinal proxy on the culture. The overt story is about boys with money in Hollywood, with the young men juxtaposed with agent Ari Gold (Jeremy

Piven) and his assistant Lloyd (Rex Lee), but all these men are revealed as boys when juxtaposed with the women who are professionals and adults. The long arc of the series appears to follow the boys' very gradual maturation into men but, when the series ends after eight seasons, it requires drastic changes to wrap up the series with most of the boy-men abruptly becoming adult males by coming into money, coming into power, or marrying, remarrying, or recommitting (with marriage on the horizon).

Californication (2007–2012) works the same territory with an older character and provides an interesting comparison. While the series minimizes star David Duchovny's age as much as possible—51 at the time of this writing—the series only underscores the theme of a no-longer-young male in Hollywood refusing to leave childhood behind.

Hank Moody (David Duchovny), a mildly drug-abusing, mildly alcoholic, mildly sex-addicted complete narcissist of a writer who is better than Los Angeles, better than the scripts he writes, and better than everyone around him, except his estranged but loving (technically ex-)wife and daughter who forgive him no matter what, because he's . . . special. (Not surprisingly, David Duchovny is one of the show's producer.)

The show is well-written, entertaining, and fascinatingly amoral. *Californication*'s overarching message seems to be a male fantasy of being able to do anything and always be forgiven by one's wife and child, Hank does abusive and self-destructive things, refuses to take responsibility for them, but is always forgiven because he's too cute and talented to stay mad at. In 2012, Hank is forced to mature a bit, as he watches his daughter date a younger version of himself, but otherwise nothing changes.

The show provides an interesting gender balance to *Weeds*, where Mary-Louise Parker's character is similarly narcissistic and forgivably cute. But it might be more productive to compare *Californication* to *Two and a Half Men*'s first eight seasons. Both series present a fictional version of the actual lives of their stars that viewers would know about from gossip shows and websites. The shows offer a PG comedy version of the X-rated lives of Duchovny and Charlie Sheen, as titillating as a studio-produced behind-the-scenes promotional video.

Regardless of other contexts, the comedy of *Californication* is all about Hank's ironically detached attitude toward the excesses that keep him from the loving embrace of his family. The show continues to suggest that Hank will turn his life around but the structural problem is that, if Hank behaves well, there is no show. As with *Entourage*, once the characters decide to act like adults, the comedy is over.

From Situation to Attitudinal Comedy

The male-centered attitudinal comedy that most fully exploits the actual change in a character is *Rescue Me* (2004–2011), a series that fits nicely into a demonstration of the development from the standard episode-based sitcom to the long-arc, season-based attitudinal comedy.

Barney Miller (1974–1982) on ABC, long the overwhelming police choice as the most realistic television show about police, was focused on the (fictional) 12th Precinct detectives' office of the NYPD. An episode-based series, wacky criminals or victims of weird crimes would come into the office and detectives would comment and interact. The show was a classic three-camera sitcom shot live in front of a studio audience, with sweetening.

The show was seen as more realistic than normal because, in a comic version of what *Hill Street Blues* had done to drama, the stories focused on the personal lives and reactions of the cops, criminals, and victims. It also carried the social realism of Barney Miller guiding a shift of detectives that featured a Polish-American detective, Wojciehowicz (Wojo), who played with and against the pre-Solidarity/Pope John Paul II Polish stereotypes with a slow but sincere character, Harris, an African American writer/detective; Yemana, a Japanese American detective; Amanguale, an emotional Puerto Rican detective; an ancient Jewish detective, Fish; and Dietrich, an overeducated detective, allowing for a range of stereotypes and comic voices. Each episode had a few subplots that allowed for quirky, absurd characters and warm emotional moments.

Barney Miller depicted an office community that was very much of the sitcom tradition, but opening up to more serious personal discussion, revelations of the emotional lives of the characters. By the end of its run, the resigned attitude of the detectives was as much the source of humor as specific jokes or situations. Nonetheless, it remained episode-based.

The Job (2001–2002), also on ABC, written by Denis Leary and Peter Tolan, follows the flawed detective Mike McNeil (Denis Leary) whose career is constantly disrupted by problems with his wife, his girlfriend, his drug use, drinking and smoking, and his attraction to fellow detective Jan Fendrick (Diane Farr). While episode-based, the series developed several long arcs.

In the first, six-episode season, *The Job* advanced the level of sitcom realism. Shot with a single camera in a documentary-influenced style, the series uses no laugh track. The situations on the show are wacky—a podiatrist harasses another podiatrist by leaving human feet outside his office, Mike spies on fellow detective Jan because he can't admit he's jealous, a suspect takes Mike hostage in a bathroom while he experiences intestinal distress, actress Liz Hurley gets stalked and ends up taking Mike to dinner to make her boyfriend jealous. Wacky complications are the key to all of Mike's relationships.

On the other hand, Mike's girlfriend is molested on the subway and he gets judgmental because she doesn't wear underwear. In another episode, the comedy of his jealousy of Jan turns into Jan confessing the emptiness of her life. An undercover operation at a massage parlor leads to the male detectives ignoring the crime and enjoying the illegal services. As with *The Days and Nights of Molly Dodd*, the humor of *The Job* increasingly arises from Mike's attitude toward the consequences of his self-destructive life, not from jokes. When overt jokes occur in the show, they lean toward the sorts of practical jokes or discovered humor associated with the job and the personalities of the character, a naturalistic humor not seen in traditional sitcoms.

Mike ruins an anger management class he is required to attend. It's a classic sitcom situation that might have its roots in Ralph Kramden's bad temper on *The Honeymooners*, but *The Job* deals with it as a real problem on the way into and out of the comic set piece where Mike baits the angry people into brawling, requiring police to intercede.

There are hints of something different in the earliest shows, as viewers watch Elizabeth Hurley making sexually suggestive comments to Mike, only to have the scene continue and it become clear that viewers heard what Mike thought he heard—or wanted to hear—rather than what she actually said. This presentation of subjective reality can be credited to the influence of *The Singing Detective* but it's most interesting for its shifting to a subjective reality and shifting out of it without ever signaling the audience when it's happening.

In all of these ways, *The Job* is an early network version of what Leary and Tolan fulfill in the FX basic cable series *Rescue Me*. As with *The Job*, Leary plays an Irish American city employee whose only clarity comes from doing his high-stress job well, while struggling with alcohol addiction, trying to quit smoking and trying to keep his family together.

The most significant addition to *Rescue Me* is the continuing theme of survivor guilt in the aftermath of 9/11. Tommy Gavin (Denis Leary) is a firefighter in New York's Company 62. Gavin is literally haunted throughout the series by the ghosts of those who died on 9/11, especially his cousin and fellow firefighter, Jimmy Keefe. Most of the rest of the characters are responding in different ways to the death of fellow firemen and relatives.

Leary's Gavin is more obviously a boy than his character in *The Job*. The men of *Rescue Me* are constructed as crude children, unable to understand women, driven by a desire to do their jobs well but, otherwise, driven by desires for tobacco, for alcohol, for drugs, for sex, or for food. The comedy arises from the male soap format of Gavin trying to do his best—which is never enough—for his family, girl friends, work, and self.

A key to the male-centered attitudinal comedies is that the men are not "manly" enough; each of the series finds it own vocabulary of unmanly childishness and unmanly "womanliness." Here Gavin is regularly confronted with the need to grow up while also being the recipient of a running string of jokes that mock him for his feminine attributes, from a womanly physique to girlish sensitivity, generally made by the women in his life.

Tony Soprano denied his guilt. Tommy Gavin is defined by his guilt. Perhaps that's the reason that over the years, unlike Tony, Gavin decides to change. He stops smoking, then he stops drinking and appears to cut back on punishing himself. He surrenders to his inevitable losses in the battle between the sexes, with his family, and with bureaucracy, and finally quits his last addiction of working fires. One might argue that Gavin's *behavior* changes more than he does but, for television, that is significant change.

Rescue Me might be seen as the television novelization of the AA Serenity Prayer: "God grant me the serenity to accept the things I cannot change; courage to change the things I can; and wisdom to know the difference."

The closest series thematically to *Rescue Me* was the quite serious TNT drama *Saving Grace* (2007–2010), about a self-destructive alcoholic Oklahoma City police detective. Grace Hanadarko (Holly Hunter) was hung over, so her brother went to the Murragh Building in her place on the morning that Tim McVeigh blew it up, and he died in her place.

Over three seasons, Grace explicitly struggles with her survivor's guilt and questions of faith. Amazingly for American television, even with a guardian angel, lovers, and family reaching out to her, Grace apparently decides it isn't enough. In an extended hallucinatory—or visionary—finale, Grace blows herself up in an act that might be self-sacrifice to stop a force of evil or might be giving in to despair. The series ends with Earl (her guardian angel) in mourning as viewers decide what happened and why.

Tommy Gavin is repeatedly tempted to his own destruction but finds reasons to survive that Grace Anadarko doesn't. Which may be another way of distinguishing attitudinal comedy from drama. But both series take their characters' despair seriously and both characters deal with their despair through irony and sarcasm and the regular playing of practical jokes. These compensating behaviors by Tommy and Grace are also compensating devices by writers to relieve the dark tone of the series. What makes this worth pondering is that one series was a serious theological drama, the other was a comedy, and both needed to add practical jokes to lighten the mood.

That might be the most telling sign of an attitudinal comedy. Series like *Rescue Me, United States of Tara, Nurse Jackie,* and *The Big C* become so serious that it becomes necessary to insert comic relief into a comedy.

Reality Situation Comedy: Documentary

Ricky Gervais and Stephen Merchant's *The Office* and Larry David's *Curb Your Enthusiasm* are two of the most critically acclaimed sitcoms of the turn of the most recent millennium, and both derive from constructing a sense of reality. The differences in their approach to reality reflect the differences between the depiction of reality on television in the United Kingdom and the United States.

The Office (2001) was created as a six-episode, three-hour comedy in the form of the BBC social documentaries about everyday people. Even more than *Mary Hartman, Mary Hartman* replicated a soap opera, *The Office* mimicked its source, appearing in exactly the same form and location—the BBC.

The series was a one-off six-chapter story about a paper company office in Slough that ends with the office being closed and the incompetent David Brent (Ricky Gervais) getting promoted to take over another location. Asked to do a second series of the show for 2003, Gervais and Merchant restored the Slough office and took away Brent's promotion. Brent instead has a boss who recognizes his incompetence and the series ends with Brent fired from his job and from the management expert lecturing side-job that he was horribly unqualified to do. Rather than do a third series, Gervais and Merchant created a two-part, 90-minute Christmas special for 2003, a "where are they now?" follow-up documentary of the sort with which BBC documentary viewers were very familiar.

The sad joke of the first series of *The Office* is that Gervais's David Brent sees being filmed for this dull documentary as his chance to show himself as the entertainer he wishes he were. (This joke is funnier in Britain, where alumni of documentaries like *Driving School* and *Airport* actually did become minor celebrities, with one—Jeremy Spake—becoming a BBC presenter [host].) The joke turns sadder in the Christmas Special, where viewers discover that everyone in public seems to recognize Brent, but only as a no-talent loser who made an ass of himself on television.

Each series tells a complete story with a different tone. The first ends with ironic absurdity as the incompetent is rewarded, and the story is over. The second series explores the consequences of Brent's incompetence and ends more painfully. The Christmas Special moves everyone's stories further along, but with a different tone as it dispenses happy endings all around, at which point the series is over for good.

By contrast, the American remake of *The Office* (2005–) was created to run forever and thrive in syndication. It uses long-arc stories more than most American sitcoms, but each episode offers closure and stand-alone pleasure, although regular viewers get more laughs from knowing the backstories of characters and events supporting the major plot of the episode.

The U.S. show also avoids change. The characters' desks might change, or they may go off to try another job or have a baby or go to art school, but their roles in the dynamics of the office family remain the same. A first-season viewer can easily follow a seventh-season episode.

To fit American network needs, the incompetent Brent is replaced by Steve Carrell's Michael Scott, who suffers the same delusions that he is a great entertainer but is quickly revealed to be surprisingly good at his job. The disturbingly, and cadaverously, wrong British general manager Gareth Keenan is smoothed out into the weird but lovable Dwight Shrute. The implied romance that never quite happens between the UK sales rep Tim and receptionist Dawn (Jim and Pam in the United States) becomes explicit and central to the American series.

The social documentary conventions of *The Office* UK have no real context in the United States, so *The Office* US devolves mainly into using the device of characters speaking directly to camera as an additional device for creating comedy. However, for most U.S. viewers, this is not a documentary reference as it is a convention of "reality" programs like *Survivor* (2000–), *Big Brother* (2000–), and the *Real Housewives* franchises (2006–), where every event is then commented on by a participant.

"Reality" Comedy: Reality Shows

Curb Your Enthusiasm (2000–present) has the same interest in the presentation of reality, but not the reality of BBC social documentary. David's sitcom arises out of the "reality" show conventions that replace documentary for most U.S. television viewers. *Curb Your Enthusiasm* draws on the idea that viewers watch "reality" shows to watch dysfunctional people go about their normal lives and laugh at them and cringe with embarrassment for them, depending upon how much human empathy survives the encounter.

Curb Your Enthusiasm is loosely scripted to create episodic plots and season-long arcs that resolve comically in the final episodes of the season. The cast is encouraged to improvise within the needs of the script, just like "reality" show casts. Some actors play fictional characters based on real people. Cheryl Hines played Larry's wife and, when he and his actual wife got divorced, so did Larry and his fictional wife. Shelley Berman, Paul Mazursky, and Vivica Fox play fictional characters. But Richard Lewis, Ted Danson, and Michael J. Fox play versions of themselves.

In Season Seven, the former cast of *Seinfeld* play versions of themselves in a season-long arc built around the creation of another final episode of *Seinfeld*, since they all agree the original one sucked. We see very different—"realer?"—interactions between the actors that we confuse with the characters, including addressing Kramer's (Michael Richards') racist meltdown at a comedy club. The TV version of Larry casts his (TV) nearly ex-wife to play the wife of George Costanza (another level of fictional version of Larry David) in an attempt to win her back, but then TV Larry gets jealous of his TV fictional self-portrait (George, played by Jason Alexander) getting too chummy with his TV near-ex-wife and sabotages the show, insisting on playing George himself (that is, a Larry David–played TV Larry playing George, a fictional version of both Larrys) before he gets thrown off the set.

The fourth season is built on the fiction of Larry David being asked by Mel Brooks to come in as the new Max Bialystock in Brooks's long-running Broadway musical hit *The Producers*. As other stories play out, viewers get regular updates of David trying to learn how to dance and sing, badly, and his increasing anxiety about being unable to perform on stage even as Brooks assures him that he'll be fine. The scenes recall the background footage that fills much of the "reality" competition shows like *Dancing with the Stars* (2005–), *American Idol* (2002–), and their ilk. The season ends with Larry on stage in an actual Broadway theater, performing and, against all odds, succeeding. At this point the camera cuts to Mel Brooks and wife Anne Bancroft in the back of the theater, cursing, because they had expected Larry David to fail so miserably the show would finally be canceled and they could get on with their lives—at which point, viewers realized that the season wasn't just a preparation of Larry to star in *The Producers*, the season was a variation on the plot of *The Producers*, theater producers planning to profit by making a guaranteed failure, only to have it go wrong because the loser they were counting on to destroy the show failed to fail.

These emotions and manipulations, the petty whining, and narcissism of the well-off are staples of celebrity "reality" show like *Denise Richards: It's Complicated* (2008–2009), *Kendra* (2009–), and *Keeping Up with the Kardashians* (2007–). A show like Bravo's *Flipping Out* (2007–) or any of the *Real Housewives* franchise shows (2006–) uses the same narrative structure—a loosely constructed plot with irritating "real people" improvising their parts within an understanding of the function of their character in the story, but arguably not as effectively. ("Arguably," because the "reality" shows are far more commercially successful than *Curb Your Enthusiasm*, even while being generally dismissed by critics.)

Larry David constructs his improvisations with skilled improvisational actors from a well-written script outline with a clearer understanding of what the show is doing. Larry David does not see his show as a way to promote his latest vanity project the way that most of the *Real Housewives* do at this point, but he deals with Jason Alexander selling his lousy vanity book on acting, he depicts the cast of *Seinfeld* as only a few steps from the cold self-interest of the *Real Housewives*, and presents himself seduced into a vanity project of starring in Mel Brooks's *The Producers*. The difference from the *Real Housewives* is the levels of comedic pay off.

Comedies have long blurred the line between the "real" comedian and the sitcom self. *I Love Lucy* where Lucille Ball played the wife to her actual husband, Cuban band leader Desi Arnaz (Ricky Ricardo on the show), Ozzie and Harriet, Abbot and Costello, and Burns and Allen all played on the confusion of sitcom role and nonsitcom celebrity to expand the reach of their comedy. In the 1970s networks began to offer successful comedians sitcoms based on their personae, a strain that starts around the time that Gabe Kaplan gets *Welcome Back, Kotter* (1975–1979) and Freddie Prinze gets *Chico and the Man* (1974–1978), and runs through Roseanne Barr, Ellen Degeneres, and Jerry Seinfeld.

All these shows were only loosely based on the actual lives of the comedians, but the shows' entertainment value was increasingly grounded in the idea that the sitcom reflected the actually life of the star. Roseanne's and Ellen DeGeneres's sitcoms, *Roseanne* and *Ellen* (1994–1998), were so grounded in the public perception of their creators that Roseanne's private life was read through events in the series and Ellen's fictional character's coming out as gay may have been a bigger media event than the real Ellen's announcement.

On *Seinfeld*, loosely based on Seinfeld's act and life, Jerry plays a fictional version of himself, Elaine is inspired by comedian Elaine Boozler, George Costanza by series co-creator Larry David, and Kramer is inspired by David's neighbor, Kenny Kramer. Many of the events and characters in the series were inspired by real events, with shout-outs to their favorite diner, Tom's Diner (appearing as Monk's), the Soup Nazi (Al Yeganeh, owner of Soup Kitchen International), the J. Peterman catalog company (with Peterman named Jacopo instead of the actual John), and George Steinbrenner (depicted as George Steinbrenner, but voiced by Larry David). The key point is not that the show was inspired by real life, but that this information was made easily available to viewers as part of the promotion of the show. The result was to "widen" the narrative in a variation on what shows like *Lost* (2004–2010) and *Heroes* (2006–2010) did with online expanded stories and activities. Viewers were essentially encouraged to follow the episodic show as the long-arc story of Jerry Seinfeld's real life through the episodes of Seinfeld and the supplemental promotional material.

Moonlighting (1985–1989), one of the claimants to being the origin of the hour-long dramedy, came to represent a different route toward reality-based comedy for reasons the creators could not have imagined.

Moonlighting had no longer arc built into the scripts, except the inevitable romance between the two leads that hated each other on first sight. Its major comic innovation was to reverse the screwball formula of overly rigid, rational man attempting to establish order in the face of a screwy dame who was more in tune with the insane world of love. On *Moonlighting*, the wacky screwball dude faces the controlling woman. Each episode offers a fairly traditional episodic television mystery, but the individual cases are unimportant; the increasingly self-referential banter between the two leads is the only event that matters. It's why the show could become an Orson Welles–narrated film noir for an episode ("The Dream Sequence Always Rings Twice") or a *Taming of the Shrew* pastiche ("Atomic Shakespeare") for another. The true subject of the show was Cybil Shepherd and Bruce Willis, and then Bruce Willis and Cybil Shepherd.

Real-life gossip supplied the long arc that turned *Moonlighting* into a longform series. As *Moonlighting* proceeded, the off-screen tension between its two leads, and between them and the producers, increased as Willis's celebrity status eclipsed Shepherd's and contract negotiations led to production breakdowns and walkouts. In the pre-Internet age, the show regularly included allusions to the actors' lives outside the fictional series, which would have been fourth-wall-breaking Brechtian alienation techniques if it weren't for the fact that there were no walls. There was no serious suspension of disbelief in the construction of the show.

At the time, the closest parallel was probably seen on *Mork & Mindy* (1978–1982), a sitcom with mediocre stories that many people sat through to see Robin Williams (Mork) step out of the plot and just riff for a couple minutes each episode. Robin Williams' stand-up career was the longform story, *Mork & Mindy*, was the way to keep updated.

More recently, the best parallels to *Moonlighting*'s strange transcendence of the show are found in Charlie Sheen on *Two and a Half Men* and David Duchovny on *Californication*, each playing a PG comedy version of the self-destructive aspects of their actual lives. Thanks to gossip sites and shows, the audience watches those shows through the lens of the continuing coverage of Sheen's exploits and Duchovny's sex addictions.

In this context, *Moonlighting* might be seen as an early statement of the appeal of "reality" programming, a comic exploitation of the same voyeuristic appeal that has launched a thousand "reality" shows. One episode of *Moonlighting* titled "The Real Poop" was hosted by gossip columnist Rona Barrett (The Perez Hilton of her day) who revealed the "real" (that is to say, "concocted for comic effect") backstage drama of the detective agency/television show including interviews with Shepherd's actual ex-husband, Peter Bogdanovich, and Pierce Brosnan of *Remington Steele* (1982–1987), a show with a very similar premise, but an entirely different tone.

Chapter Nine

Graphic Novels and DVD Novels

In Which We Consider Comics and Comic Book Stories, Graphic Novel Storytelling, and the Increasingly Permeable Boundaries between Different Forms of Media

It is not a genre. It is simply a medium. And it's a medium that you can do anything with. I guess because stories drive me, what then fascinates me is what kinds of stories can you tell.

—Neil Gaiman[1]

Comic Book Television

Television series have been based on comic books since, at least, the 1960s, with many being transferred fairly directly into Saturday morning cartoons featuring DC and Marvel superheroes or spin-off characters from *Archie* comics. Later, animated adaptations of comic book superheroes adjusted to an older, ironic audience with series like CBS's *Teenage Mutant Ninja Turtles* (1987–1996) and Fox's *The Tick* (1994). There was also the stylish animated *Batman: The Animated Series* (1992–1998) that developed at the end of the century, episode-based but with some longer direct-to-DVD "movies." Live action series have also adapted comic book stories. *The Adventures of Superman* (1952–1958), *Lois & Clark* (1993–1997), *Smallville* (2001–2011), *Sabrina, the Teenage Witch* (1996–2003), *The Middleman* (2008), *The Human Target* (2010–), and the tongue-in-cheek *Batman* (1966–1968) that actually used multi-episode cliff-hanging stories. Live action anthology series based on comic books have included *The Incredible Hulk* (1977–1982) and HBO's *Tales from the Crypt* (1989–1996).

But a new sort of comics-inspired series emerged around the turn of the century. Inspired by the narratives of longform comics/graphic novels, shows like *Heroes* (2006–2010), *Lost* (2004–2010), and *Carnivale* (2003–2005) were developed. The influence of graphic novels increased as viewers could increasingly be expected to watch the screen; shows such as *Rubicon* (2010), *The Killing* (2010–), and *Breaking Bad* (2008–2012) incorporate the visual language and pacing of minimalist graphic novels. Among the various graphic novels optioned for television development, *The Walking Dead* (2010–) broke through to become the first longform television series that adapted both the story and style of graphic novels to television.

While I recognize the aesthetic bigotry inherent in this subjective judgment, the term "comic books" will be used to refer to freestanding issues and to genre comic book stories, "graphic novels" will be used to refer to longform, novel-length stories (generally the republication of a series of individual issue "chapters" in a paperback book) with broader artistic aspirations, and "comics" will be used to refer to both collectively.

Graphic Novels and DVD Novels

The relationship of the comic book and the graphic novel is a lot like that of broadcast television and the DVD novel. Comic books are short, ephemeral episodic stories. Graphic novels are compilations of the individual stories of comic books that present the individual issues as chapters in a longer narrative.

The rise of the graphic novel made comics stories more permanent. Bound between firmer covers than comic books, they also became socially acceptable and worthy to be shelved in bookstores and public libraries, and reviewed as literature.

The existence of the graphic novel also changed the sorts of stories that got written in comics. Direct-to-graphic-novels became more common.

The DVD novel has developed in the footsteps of the graphic novel. Television-season DVD sets have changed the sorts of stories made on television, are increasingly shelved in libraries, and are now taken seriously enough to be reviewed next to films on DVD. So far, costs have hindered the arrival of the direct-to-DVD novel.

Comic Book Writing

In Chapter 8 (on British television), the convention of novelists writing television was discussed. The American equivalent seems to be the increasing number of comic book writers making television. The demands of episodic writing remain the same. Comic book writers bring a strong awareness of working within genre conventions from the superhero franchises but also an awareness of the narrative experimentation of smaller, auteur-driven comic book series that reach their long-term audiences in the graphic novel format. Both types of comics are aimed at the same demographic as most television shows.

A comic book series like *Lost* had a number of comics writers working on the show, including Javier Grillo-Marxuach (the creator of *The Middleman*, a comic series that was adapted to television by ABC-Family), Drew Goddard (a Joss Whedon alum who wrote part of the *Buffy the Vampire Slayer: Season Eight* comics), and Brian K. Vaughn (whose *Y: The Last Man* and *Ex Machina* have attracted interest as potential television series). Jeff Loeb came out of genre film writing, wrote for the teen Superman series, *Smallville*, and then became a supervising producer on *Lost* before writing and executive producing *Heroes*. Throughout that time he was also writing superhero comics for Marvel and DC.

A previous generation of writers who moved from comics to TV includes Gerry Conway (who killed Gwen Stacy and created the Punisher for *Spider-Man* in the 1970s) has written and produced for the *Law and Order* franchises (including the Russian version of *Law & Order: Criminal Intent*). Len Wein who created *Wolverine* writes *Ben 10*; Steve Gerber (*Howard the Duck*) was story editor for *GI Joe* and *Transformers*, Ben Edlund who created the *The Tick* worked on the show and has since become a regular of Joss Whedon's stable of writers for TV. DC Comics' Chief Creative Officer Geoff Johns has regularly moved between television and comics.

One of the main writers for the non–comic book series *Big Love* (2006–2011) was Roberto Aguirre-Sacas, while he was also writing superhero comics for Marvel and establishing himself as an award-winning playwright. In the last decade, Aguirre-Sachs adapted Stephen King's *The Stand* to comic books, joined the writing staff of *Glee* (2009), and successfully rewrote the book for the previously disastrous Broadway musical *Spider-man: Turn Off the Dark*.

Heroes

The most obviously comic book narrative series is *Heroes*, the serial story of normal people who discover their superpowers and use them to save the world, while remaining fairly normal.

Creator Tim Kring modeled the show on the storytelling of comic books, developing smaller stories within each long story arc (referred to as a "volume"). In the four seasons of *Heroes*, there were five full volumes—Genesis, Generations, Villains, Fugitives, and Redemption—and an uncompleted sixth volume that began as the series ended. Each volume is a complete DVD novel. Each new volume begins as a sequel, after time had passed from the previous volume. With the second season, ratings dropped amid complaints that the show was slowing down and was too focused on plot and character development to the detriment of action.

Heroes may be most significant for its extensive attempts to expand its narrative presence across media, including self-parodies on YouTube, video games, and a weekly web comic (eventually assembled and published as a DC graphic novel). There were also *Heroes* novels covering adventures not occurring on the show: online manhunts, games, trivia, and a *Heroes: Evolutions* site that expanded the narrative. In an expansion of the attempt to give viewers what they want, NBC (sponsored by Sprint) added a "Create your Hero" in which viewers voted on attributes for a new hero to be created and featured on a web-based series, *Heroes: Destiny,* at the end of 2008. NBC had agreed to a spin-off series, *Heroes: Origins*, beginning in 2007 but it was canceled due to the Writers' Guild Strike.

This comic book industry approach to market expansion is a significant influence in expanding not just the length of television narrative but also the "width." DVD sets become part of a strategy of "wide-form" television, delivering additional stories, deeper backstories, and increased viewer interaction beyond the television broadcast.

In this, *Heroes* built on similar expansion by *Lost*, including novels, an online diary of Janelle Granger, *Lost: Missing Pieces* "mobi-sodes" created for cell phone viewing that eventually appeared on the fourth season DVD set. *Lost*'s online presence included more esoteric faux websites like one for Oceanic Airlines and another for the Hanso Foundation. Online games include the 2006 ARM game online, *Lost Experience.*

Central to the "width" of *Lost*'s narrative was the idea that just watching the show was not enough. The promise of deleted scenes and extra content on DVD sets made them top sellers upon release, with estimates as high as half a million sales on the first day of release for the "enhanced" Season Two set. The centrality of the DVD set to *Lost* was confirmed when the show ended and the 12-minute epilogue, "The New Man in Charge," was only available on the final season DVD set.

Graphic Novels as Source Material

The advantage of adapting a successful serial story, whether graphic or traditional novels, is that the material has had the opportunity to develop a following and run long enough that producers will know how the story will play out. Comics offer the added advantage of being a set of storyboards for the television production.

A less frequently discussed advantage to adapting comics to television is that they are easier to translate than the average novel. One of the difficulties of adapting traditional novels to television is that a triple-spaced page of a script is generally calculated as a minute of screen time. Adapting a 300-page single-spaced novel becomes an exercise in cutting away.

The comics narrative is generally already stripped down by the limited space for words on the comic page and the conventions of visual storytelling. The form even leaves the adapters the space to add.

Comparison of *Game of Thrones* (2011–) to *The Walking Dead* adds another level to the differences in adapting traditional and graphic novels. The work of George R. R. Martin has translated to a successful HBO series that will undoubtedly become an even more successful series of DVD sets. But the strength of Martin's novels, beyond the extended plotting, is the complicated intrigues of fully rounded characters written from their subjective points of view. On television, almost all their thinking and feelings are lost as the 720-page *Game of Thrones* must be stripped down to its 11-hour essence. By contrast, *The Walking Dead* graphic novels leave the viewers to discern emotions from visuals and minimal dialogue, and leave the television writers space to flesh out the characters with the comparative luxury of added dialogue. Ironically, viewers responded badly to a dialogue-heavy third episode of the series' second season, for telling rather than showing.

There may be something about the "thinner word, thicker visual" form of the graphic novel that makes it naturally more successfully adaptable to the televisual form.

The Walking Dead (comics started 2003, television 2010) is the most successful example so far of a comic book turned graphic novel adapted directly into a television program turned DVD novel. Film director Frank Darabondt's first season and a half of the series stayed close to the plot and visual storytelling of Robert Kirkman's graphic novel as drawn by Tony Moore for the first seven issues and Charlie Adlard after that. The visuals establish the slow empty reality of the world, which is reinforced by dialogue and action. While each episode of the story offers the closure of the survival of a zombie attack, a human lost, or a lesson learned, the series focuses more on the mundane elements of life rendered epic in a postapocalyptic world. Both the graphic novels and television adaptation benefit from their episodic nature, because the inevitable end of the series is that all die or become zombies but the closure of the separate episode protects the audience from seeing that all of the characters, and humanity, are doomed.

The main obstacle to adapting the most popular graphic novels to television has been the perceived costs of adapting supernatural events for the smaller screen. For years, there has been talk of adapting Neil Gaiman's epic 10-volume graphic novel series, *The Sandman*, to television, preceded by many years of speculative scripts for a film adaptation of part of the first book in the series. Joe Hill's *Locke & Key* (2008–) hit horror/fantasy comics series had a pilot made but the series was not picked up. The pilot has been a big hit at convention screenings but, to date, the budget has been seen as prohibitive for a horror/fantasy series not involving lusty vampires.

ABC's *Once Upon a Time* (2011–) owes more than can be legally acknowledged to the epic story told by Bill Willingham's (so far) 15 graphic novels in the *Fables* (2002–) series. The series which is the story of trans-dimensional war that divides the entire universe of fairy tale characters between those hiding among normal humans in New York City and their allies, and those fighting for or living under the Adversary, has been Vertigo's highest earning franchise. Having completed the epic war and its consequences, the series moved on to individual graphic novels recounting the adventures of specific fairy tale characters in the mundane world.

Fables was optioned by ABC in 2008 but the production wasn't picked up. *Once Upon a Time* is credited to Edward Kitsis and Adam Horowitz, executive producers of *Lost* who had also created the unsuccessful DC comics superhero series, *Birds of Prey* (2002–2003). One of *Lost*'s co-creators, Damon Lindelof, consulted on the development of the series, where fairy tales characters have been transported to a New England town and stripped of their memories, but conflict from their fairy tale past continues to intrude into the mundane world. As with *Fables*, part of the viewer pleasure comes from determining the fairy-tale identities of supporting characters.

Television Visuals and Comics Language

As seen with *The Walking Dead*, one way that comics narrative has influenced longform television is in the use strategies of information withheld. While it is possible to have a character's thoughts or monologue spelled out on the page for a whole issue of a comic, it's boring. The visual is half of the equation. Often more. The form leads naturally to a reliance on the reader to gather meaning from what is not said and what is seen.

Even though he is not an artist, Stan Lee is often credited with changing the way comics were drawn in the 1960s. Where earlier superhero comics had less dynamic visuals, Marvel comics under Lee adopted what was called a "cinematic" style of foreshortening and exaggeration of heightened perspective. This cinematic style was a dramatic use of the visual space that went where films would eventually go, but had not actually arrived.

Visuals ceased to be the servant of words in American comics storytelling in the last third of the 20th century, a fundamental shift from a simply illustrated book. Underground comix took visual-based narratives even further. As television screens became bigger and incorporated comics visuals, a similar shift occurred on television.

It is interesting that the use of visually oriented television is called cinematic. It is true that, from the 1980s, television series were increasingly trying to get the film look on a television budget. But a couple decades down the road, the visuals of comics and video games have developed original visual languages that have come to shape television and film.

Director Zack Snyder provides an easy example of comics' visual storytelling because he made two films that explicitly copy two very different comic book styles. Snyder's adaptation of Frank Miller's *300* (2006) and Alan Moore's Dave Gibbons–drawn *Watchmen* (2009) use the panels of the original comics as storyboards for the live action films. In both cases, the power of the film is in the visuals. *300* uses simple but color images derived from the minimal black-and-white images of Miller's comics. By contrast, *Watchmen* repeats the dense images of Gibbon's panels, with stylized color highlighting background imagery that carries large amounts of information, including hints about relationships, history, and character backstories, but also foreshadowing plot and larger themes that will eventually emerge. *Watchmen* requires multiple viewings to appreciate the depth of its visual storytelling.

For some, this is a cinematic use of the film visuals—communicating meaning through the mise-en-scène—but the amount of visual detail exceeds the ability of a casual film viewer to absorb it. A comics reader—or a graphic novel viewer—can take as long as they want to ponder each page or shot.

Sophisticated visual storytelling is increasingly common in longform television, perhaps simply because it is expected that viewers will actually *watch* the shows on DVD sets. Traditional television narrative began as, and for many remains, "radio with pictures." Television writing initially required a word-based storytelling accompanied by simple visuals because of small screens and unreliable reception.

Even after the shift to more cinematic television in the 1980s, network television continued the emphasis on words in the assumption that audiences don't actually *watch* television, they monitor it while doing a variety of other tasks. The failure of the Zucker Brothers' *Police Squad* (1982) can be attributed to its airing in a time slot when people were not actually viewing, but listening, thus missing the series' sight gags.

The distinction between cinematic television and comics visuals on television is subjective. *Lost* and *Heroes* both work with a comic book style of narrative using fairly traditional television visuals, even though *Heroes* self-referentially uses comic books and comic book panel paintings within the story. In some ways, this is splitting hairs; a long shot of empty desert with wind blowing across it and only the shadow of a person coming in from the foreground, with no dialogue. Is that cinematic or comics?

One way to resolve the distinction is to say that it is cinematic language that gets used more frequently in graphic novels than it does in Hollywood films. Comics have absorbed and updated the visual vocabulary of silent film, with dialogue balloons taking the place of the old interstitial title cards. In general, non-superhero graphic novels also use simpler graphics—often black-and-white visuals—and sparser dialogue, with more emphasis on mood than action, with visuals being an equal if not dominant partner with words.

The increasing use of the still shot would be one specific graphic novel influence on television storytelling. A single, unmoving close-up shot of a plant or an eyeball or a gun, or a locked-off camera shot with a simple motion within the frame—a turtle with a decapitated head on its back walking slowly through the shot, a bulldozer slowly moving into a shot of a house on a flat horizon, or a cop sitting in a car in a driving rain with traffic blurring past outside. These are all shots that ask television viewers to contemplate the image in a similar way to the demands of certain comics panels.

Similarly, the vocabulary of juxtaposed, still, apparently unrelated shots seen at the beginning of the third episode of the second season of *The Walking Dead*, later explained through expanded shots at the end, is far more a convention of the graphic novel than any American film style.

Graphic Novel Storytelling in Nonadapted Television

Carnivale, The Killing, Rubicon, and *Breaking Bad* are all series not based on comics that, nonetheless, share the visually based storytelling and grand sweep of graphic novels. *Carnivale* uses not only the visual style, but the epic storytelling of vast historical epics and battles between good and evil seen in many graphic novels.

HBO's *Carnivale* (2003–2005) is the tale of a young man traveling with a carnival during the Great Depression, who gradually discovers he has healing powers that put him at one end of a multigenerational battle to the death between forces of good and evil. The series was designed as a six-season trilogy, with each "novel" made up of 24 chapters over two seasons. The series would recount three generations of the continuing struggle between opposing supernatural avatars.

On the one hand, the series was created by Daniel Knauf, a writer of genre television, film, and, following *Carnivale*'s cancellation, comic books for Marvel. On the other hand, the visual style of the series was largely created by Rodrigo Garcia, the director of the first episodes of the series. Besides being a writer and director of his own films (*Nine Lives*, *Mother and Child*), Garcia is the son of Gabriel Garcia Márquez, noted Colombian author of *One Hundred Years of Solitude* and *Love in the Time of Cholera*. He grew up in a household of collaborative art where he learned that images were as important words.[2]

The meeting of director and material was especially fortuitous as the series drew on symbols without clear meaning, visions from dreams, the day-to-day world of carnival freaks, and the everyday existence of supernatural events. Garcia's childhood exposure to the blended logics of magical realism made him at home with stories simultaneously mundane and mythic.

Key moments in the series are communicated in visuals, not words, often played out against the vast emptiness of a 1930s Dust Bowl landscape that reads like the horizon line in a comic book panel. The juxtaposition of small people with large indifferent landscape may draw on memories of WPA photography and John Ford's *The Grapes of Wrath*, but it is also contrasted with the congestion, if not claustrophobia, of the carnival at night and the private spaces of the company. Clancy Brown's Brother Justin Crowe, the eventual antagonist to Oklahoma farmboy Ben Hawkins (Nick Stahl), is visually contrasted with Hawkins—tall, erect, dressed in black in opposition to Hawkins's scrawny hunched and dusty pale form, just as his settled world is a visual contrast to the world of the carnies.

Carnivale's casting provides the series with characters that are visual "types," communicating immediately the generic roles they serve. Hawkins is a damaged boy trying to be a hero, Brother Crowe is a too-powerful man busting out of the moral limits of his ministry, Jonesy (Tim DeKay) is a noble but not too smart guy, and Sofie is an abused woman who would be beautiful if she came into her own.

It is also a world of external storytelling, where we watch characters from the outside who occasionally explain themselves, but information is just as likely to be found in a close-up that dwells on a letter on a desk, a handbill on a wall, or a symbol we don't understand but recognize has significance that will eventually be demonstrated and, perhaps, explained. When Hawkins uses his power to give sight to a blind girl, the camera slowly pans across acres of crops dying as we watch, teaching us about the price of the show's miracles.

The nominal romantic interest, Sofie (Clea DuVall), hears her comatose mother's fortune telling and warnings, but viewers learn that only from Sofie's side of an otherwise silent conversation. Samson (played by Michael J. Anderson, the backward-talking dancing dwarf of *Twin Peak*'s visions) engages in similar one-sided conversations with the monstrous Management and withholds more than he reveals as he guides Hawkins toward his destiny.

AMC's *The Killing* is the remake of a Danish series, *Forbrydelsen* (2007, 2009, 2012). The first season of both shows tells the story of a police detective who discovers a girl's murder on her last day at work, a murder that seems connected to a mayoral election. Comparing the American remake to the original Danish series, the minimalist style choices made by the Americans become clear.

The American version strips away the visual sensuality and most of the engaging camera movement and edits. Images are grayed and slightly blurred. The constant, even overlapping, dialogue of the original gives way to long near-silent scenes and shots in which nothing seems to happen. Where *Forbrydelsen* places characters in a space with depth and saturated color, populated with visually interesting people and objects, *The Killing* is more likely to shoot characters against spare or minimal backgrounds that suggest isolated individuals in empty, if shallow, space.

The two series demonstrate a cinematic and a graphic novel visual language. The much-criticized minimalism of *The Killing* is only a more extreme version of the stripped down aesthetic of graphic novels like *A History of Violence* and *The Walking Dead.* And their televisual adaptations.

This can also be seen in *Breaking Bad,* which often uses long-held shots in which not much happens or what does happen is not clear. Information is regularly withheld, with sparse or banal dialogue and images that viewers must interpret. The audience is forced to settle into the pace of the story and, more often than not, mislead themselves about what matters. When activity and genuine dialogue occur they recontextualize what viewers have seen up to that point and start the audience on another arc of interpretation and mood.

This is a form of storytelling that rewards patience and close attention, neither of which are traits that traditional television shows encourage.

Television to Comics

One final demonstration of the relationship between television and comics narrative is the often unfulfilled promise of a canceled television series continuing in comics. This solution probably makes more sense than the more expensive unfulfilled dream of continuing a canceled series as a feature film (Veronica Mars joins the FBI), or a series of feature films (*Deadwood*). Television stories are just more similar to comics than they are to film.

After *Buffy the Vampire Slayer* (1997–2003) ended, Joss Whedon continued the series into Season Eight as a Dark Horse comic book series, republished as a set of graphic novels. Buffy, Willow, and the gang move on to a global adventure that includes time travel and superpowers, with expansive sets, special effects, and a large cast of characters beyond the budget of any television series. Season Eight apparently got all of the big budget story out of Whedon's system and Season Nine has been announced with the promise to ditch the spectacle and return to the show's character-driven roots. Whedon also uses graphic novels to continue *Firefly* (2002) using comics and original graphic novels to fill in backstories for the characters and continue their adventures.

The main hurdle in continuing a longform television series in longform comics format has been the issue of getting permission from whoever legally owns the rights to the characters, the series, and the title for nontelevision development, which is not always the creator. When Whedon made the *Firefly* into a film, he titled it *Serenity* (2005) to avoid testing Fox's claim of ownership of the series title for film. But Whedon has generally maintained the rights to his stories to an extent that other creators have not.

When *Carnivale* was canceled by HBO after the second season/first volume of the trilogy, fans, Knauf and, apparently, Marvel comics agreed that the series would be a natural to be completed in comic book form.

Unfortunately, HBO didn't agree, so Knauf moved on to writing other comics for Marvel and fans of the show need to settle for commentary tracks and blogs to discover how the last two-thirds of the story would have gone. Apparently the same thing happened with the attempt to continue Veronica Mars as a longform comic series; there was much enthusiasm from fans and series creator, Rob Thomas, a DC comics fourth season of *Veronica Mars* was announced in 2007, but apparently ownership questions got in the way.

Chapter Ten

David Milch and Television as Literature

In Which We Consider Magical Realist Television, the Authorial Voice on Television, the Heroic Community, and the Way the Meanings of Language Lead to the Meaning of Meaning When William James Meets 12-Step Spirituality on a Long Strange Trip from Harvard to Deadwood and the Beaches of Southern California by Way of Rome

All my stories are about the action of grace on a character who is not very willing to support it, but most people think of these stories as hard, hopeless and brutal.

—Flannery O'Connor[1]

David Milch provides an interesting contrast to Bochco. Both men push the television form but they do so from different backgrounds and with very different results. Bochco trained to make network television. Milch trained to write and teach literature, while plumbing the worlds of gambling and addiction, and then ended up working with Bochco.

Milch seems a teller of tales in the grand tradition of barroom storytellers and junkie con-men more than the academic, although the two strains repeatedly combine in his work. He is the most literary of successful American television writers, drawing on his own life as the inspiration for series that explore issues of human interaction and spirituality quite beyond what one expects from television, but also using television to teach—to dissect the meaning of symbols, the act of communication itself, and to preach the connection between all humanity.

Milch writes for the ear but for specificity as well. Where David E. Kelley writes his shows in longhand on yellow legal pads, where everyone has similar absurd banter, Milch takes the next step and dictates his scripts to assistants as an audience, playing every distinct role.

Milch is a hard man to avoid analyzing through his biography. His basic storytelling seems to grow out of his life story, raised by an alcoholic doctor-father who introduced him to gambling and gamblers, developing an ear for dialogue that Robert Penn Warren famously praised as better than Hemingway's. Milch was on track to be a novelist and an academic. He was already a heavy drinker by the time he arrived at Yale, and a gambler with obsessive-compulsive tendencies. Soon he had added heroin to the mix, supposedly enhancing his reality with the drug through most of his teaching and television writing careers, including a start on *Hill Street Blues* (1981–1987) and then his running of *NYPD Blue* (1993–2005). With *Deadwood* (2004–2006) Milch cleaned up and began to concentrate on using television writing as a medium for discussing and exploring the deepest social and spiritual issues.

This is not the personality, or the addiction, that one expects to find in television in Hollywood, where cocaine is the drug of convenience in the long hours of writing. But Milch's roommate from Yale, Jeffrey Lewis, was writing for *Hill Street Blues* and invited Milch to join him. His first script won a plethora of awards, impressed Bochco with Milch's genius, and the academic's career as a television writer was under way. After *Hill Street Blues*, Milch created a couple of unsuccessful shows before he and Bochco created *NYPD Blue* and Milch went on to shape the character of Andy Sipowicz out of his past—what he described to Mark Singer in a *New Yorker* profile as informed by the guiding influence of Warren's poem, "I Am Dreaming of a White Christmas: The Natural History of Vision" in seeing the way that "pain of the past" may become "the future tense of joy."[2]

NYPD Blue

NYPD Blue gives audiences a seriously unattractive hero in Andy Sipowicz, an Archie Bunker unfiltered, with the rough edges unsoftened by malapropisms and laughter. The series begins with him as a bigoted, alcoholic, raging detective who is about to put the last nails in the coffin of his career when he is shot and ends up forced to dry out as he recovers from his near-death experience.

Over the run of the series, Sipowicz transforms himself into the good man that was always there under the alcohol and rage. While this repeats the traditional network pattern of making unattractive characters lovable, this character was much more complicated and realistic than audiences have been trained to expect. Sipowicz's gradual confrontation of his racism forces viewers of the show to see the issue not in the television convention of injecting a racist that our proxy could condemn, but forcing viewers to feel the weight of the racism in the character through whom they have entered the show. Similarly, Sipowicz's recovery is slow and painful and not embraced by everyone. Even though he gets the girl and reconnects with his family, there are slippages and costs. But the arcing story of Sipowicz's redemption is the slowly developing core of most of the series.

Expanding on *Hill Street Blues*' realism, *NYPD Blue* replaced the cool, tempered Frank Furillo, recovered alcoholic, with a sloppier reality of recovery. It is hard to avoid reading Sipowicz as an examination of himself through his surgeon father, a network adaptation of the self-evisceration through proxy that Dennis Potter did in *The Singing Detective*. Milch does not directly address the point, simply acknowledging his father as the inspiration for the story of a bigoted, angry alcoholic who redeems himself through connection with his partners, his wife, his estranged son, and then his newborn son. Sipowicz survives crises largely out of his responsibility for and the love of his son, and he seems an emotionally true variation on Milch's father, seemingly creating a happy ending for his father that he never completely achieved in life. But Sipowicz's redemption also serves as a more distanced meditation on Milch's evolving ideas about redemption. Regardless, character transformation was not part of the story vocabulary of television at the time.

Actually, neither was the vocabulary of obscenity that Milch fought for the right to use to enhance the realism of the show. There was an initial boycott of the show by about a quarter of the network affiliate stations that refused to air the show due to its "indecency," but the popularity of *NYPD Blue* soon broke the boycott and ABC viewers across the country could hear words such as "asshole" and "son of a bitch," and see occasional moments of full backal nudity.

Besides the personal realism from Milch and a novelistic extended arc of character transformation that was unseen in American television at the time, *NYPD Blue* was still an episodic cop show with smaller arcs built around Sipowicz and his fellow cops. An increasingly central voice in the show became Bill Clark, a veteran New York police officer who eventually became a producer of the show.

At the end of the series, Milch was financially set for life, even with his extravagant gambling, and he ended up finally detoxing from his heroin addiction. After a couple years sober, he went back to network television to pitch NBC a western series, but they would only agree to fund a pilot, not a full thirteen weeks, so the deal fell apart. It's unclear how that show was to be constructed, but in it were one set of roots for the HBO series *Deadwood*.

From college, Milch was on a trajectory to be a serious novelist. At Yale, he worked for and with Robert Penn Warren, the novelist and poet who won Pulitzer prizes for both, who is generally seen as one of the great American novelists for one of his 10 novels, *All the King's Men.* The novel applied Warren's observations of the rise of fascism in Italy to the story of the rise of a small-town Louisiana politician, Willie Stark, to national prominence and corruption, narrated by a young historian who helps him advance and gradually realizes his own culpability as Stark moves from savior to monster in his eyes. The novel's story becomes the framework for a deep exploration of the sweep of American history, the loss of innocence, and the nature of humanity.

While Milch is more than an academic writer or follower of Warren, the man's influence cannot be overstated. Milch worked with and for a novelist-poet and academic who thought and cared about the ideas behind literature and the way that society and humanity are affected by art. Warren embraced popular culture themes within his writing and saw art as the place to explore ideas while also telling a story. Milch's series *Deadwood* is the work that seems most clearly related to the work of Robert Penn Warren.

Deadwood: Milch's Historical America

David Milch did not want to make *Deadwood.*

In his discussions of the themes of his post–*NYPD Blue* work, Milch seems to combine William James's ideas of human consciousness with the "higher power" vocabulary more familiar from 12-step programs, but his themes remain consistent: the impossibility of the individual alone and the necessity of community. These are two ideas drawn from both sources and they are ideas that contradict the essential Hollywood story of the solitary hero.

Milch, after *NYPD Blue,* wanted to do something more with his art than create another emotion-generating machine of a story. He wanted to explore more complicated ideas about human civilization and salvation than had been seen on television and, one suspects, he wanted to teach these ideas to a wide audience. (Not in the sense of an L. Ron Hubbard, a writer who decided to market a set of religious truths, but in the sense of the teacher of writing, literature, and poetry that he was, and remains, through seminars on weblogs.)

He settled into the idea of writing a television show set in the fall of the Roman Empire, with the arrival of Paul and the idea of Christianity—or, as Milch puts it in his lecture to a Religion and Media class, "the illusion agreed upon" of the Cross. His intention was to explore that symbol, and the attachment and meaning of the symbol, as Rome collapsed and Christianity rose.[3]

HBO said they loved the show but had already committed to another Roman show, *Rome* (2005–2007)—which became a very successful series for HBO, staking out its territory somewhere between the BBC's *I, Claudius* (1976) and a tamer Robert Guccione's *Caligula* (1979), a very popular and entertaining series with no big ideas but a running soap opera of historical celebrities with nudity, violence, acceptable images of degeneracy, and working-class soldiers with whom to identify.

So—perhaps in a subtle homage to William Jennings Bryan—Milch replaced the Cross with the "illusion agreed upon" of Gold, and created *Deadwood* (2004–2006), a series he discussed as being about the idea of gold in its absence, the way that men and society are shaped by the idea, with the intention that the series would follow the characters and their cultures—and the idea of gold—into the period when they actually acquired it, and the complications of ideas and reality, side by side.

Deadwood was an actual town in 1877, just beyond the Dakota Territory and therefore, outside the United States of America and its federal jurisdictions. The main characters of the series are fictional versions of actual people who settled the town, although the series does not feel like a history lesson because the actual people have been largely lost from a popular sense of history.

Deadwood emerges as a very personal project for Milch, a story with the same epic potential as Warren's *All the King's Men*. Milch spent two years with his obsessive impulses channeled into research of the historical and geographical region, any elements of culture that might come into play when he started writing. Before the series began he spent weeks doing seminars on the themes, history, background, and structure of the series. Once the series was under way, scripts were written or rewritten by Milch into the profane and genteel poetry he developed to represent the era. Milch was also a constant presence on the set during production. And whether he wrote the original scripts or they were written by staff writers, they passed back through Milch's monologist rewrites before they were distributed.

Each season of the show is constructed around a thematic issue within an overarching narrative arc of the rise of Deadwood as an actual town, on the way to being a part of the United States of America. The overarching theme of the first season is the clumsy attempt to get a town started. The second season is based on the fear that the gold had run out. In the third season, the possibility of gold attracts the attention and intervention of powerful men outside of Deadwood. Future seasons would have explored the community dealing with the presence of gold, with attempts to join the United States—another "illusion agreed upon"—and the shift from frontier community to established member of the "civilized" world.

The Heroic Community

Milch performs a complicated adjustment to the ideology of television drama. The convention of American television, growing out of American film, is that most stories direct viewers to follow the story of one hero, and everything that happens tells us about the character; other characters tell us how to understand the hero. Antiheroes complicated the emotions associated with the viewers identifying with the protagonist, but, as bad as they might be, Tony Soprano, Dexter, and Walter White of *Breaking Bad* (2008–2012) are still the individuals around whom their series revolve.

Ensemble series like *Hill Street Blues* complicated the relationship by giving viewers multiple heroes with whom they could choose to identify, but Milch takes it further in *Deadwood*. Milch has said that a television writer is "asked to presuppose otherness" but the stories he wants to tell say that we are not "others," we are of the same spirit.[4] So Milch's story of the West rejects the individual cowboy hero audiences are trained to expect. The hero of *Deadwood* is the whole community; Al Swearengen, the nominal villain, cannot make Deadwood viable on his own, but neither can the nominal hero, Seth Bullock. Every good and bad person in town is necessary. It takes a disparate frontier community to make a village.

In the fifth episode of the series, Milch articulates his central theme at the funeral of Wild Bill Hickok, as the Reverend H. W. Smith preaches on the text of the First Epistle to the Corinthians, Chapter 12:

> St Paul tells us "By one spirit are we all baptized into one body, whether we be Jew or Gentile, bond or free, and have all been made to drink into one spirit. For the body is not one member but many." He tells us "the eye cannot say unto the hand 'I have no need of thee.' Nor again, the head to the feet, "I have no need of thee." They, much more those members of the body that seem to be more feeble and those members of the body that we think of as more honorable, all are necessary." He says

that "there should be no schism in the body, that the members should have the same care, one to another, and whether one member suffer, all the members suffer with it.[5]

First Corinthians actually begins with Paul's more overt declaration that every person has been given a manifestation of the divine spirit and that all are necessary, although Milch is making art, not directly lecturing his audience, so he avoids that more literal statement to stress the allegorical story, making his point but also creating a meaningful comic scene where the nominal hero, Bullock, seems irritable and unable to grasp that he is only one part of the community, caviling with Sol Star about which body part he might be.

What Milch develops in *Deadwood,* a fundamental rejection of the heroic ideal of American television, resonates with what David Simon develops in *The Wire* (2002–2008) although the two come from fundamentally different philosophical and aesthetic orientations. *The Wire*, and later *Treme* (2010–), looks at the community from a journalistic vantage translated to fiction, within the Greek tragic vision of humans being the playthings of the Gods. Milch pursues a literary strategy that explores his vision of humanity's shared connection through the divine spirit. As Milch told a university class, "[You are] part of that which created you. Despite seeming separateness, the soul of the universe radiates from us all. To experience both is the state of grace, the state of art."[6]

The Language of *Deadwood*

There is also the issue of language.

The longform of television allowed Milch to develop character and explore the implications of language over time and changing context.

Milch makes the case in his *New Yorker* interview that language at its base is an exchange of energy, that meaning exists in how things are said—an idea he explored in its barest form in a first season scene where Wu and Swearengen have a conversation that largely consists of the word "cocksucker" repeated with vehement variation to communicate much more. In *Deadwood*, obscenity becomes a most basic unit of speech.

But Milch explores the forms as well as the essence of language. To call his language Shakespearean in *Deadwood*, as is often the case, is to use the term the same way that "Greensleeves" is used to provide period music for any English period from the Dark Ages to early Victorian. It's a term that reflects the paucity of modern American English, more often than not, rather than a specific quality of speech.

Deadwood's characters have heard or read their Shakespeare, they've also heard their King James Bible and listened to the poets, rather than the television and radio. The language of *Deadwood* is a language of people who weigh the value of words, on one hand, and the shape and rhythm of words on the other.

That tradition largely disappears from American writing, and conversation, around the era when F. Scott Fitzgerald's writing for the ear (and stories read out loud) is replaced by Ernest Hemingway, and the writing for the eye, the watershed of functional language in American society. As content follows form, it is probably not surprising that Milch dictated/performed his scripts to assistants who transcribed and polished them.

The anachronistic use of obscenities as an art form of the frontiersman was what first caught the ear and imagination—and the outrage—of the viewing audience, but Milch goes much further. Each character has a pattern of speech that reveals his or her character; where one's words are poured out like honey, to fill in gaps in the logic and sweet talk the foolish, another's are measured out in a mercantile—or, perhaps, hermitic—manner. Others speak the language of reporting, the language of lawyers, the language of school marms, the language of privilege, of the city street (transplanted), of the church, or of the whorehouses. And as they do, they reveal their agendas in form as much as in what they actually say.

Robert Penn Warren famously is said to have claimed Milch was a better dialogue writer than Hemingway,[7] and what might get lost is the way his dialogue is made up of very different languages. Or stylized versions of idiosyncratic transformations of the same language. At some level, all of Milch's dialogue comes down to the regular pidgin conversations between Swearengen and the Chinese boss, where each refuses to learn the language of the other, but both can understand enough from the energy of their exchanges to negotiate, to threaten, and to make a human connection.

This is a style of writing very rare in American popular narrative. While some writers attend to the general mesh of language—as *Veronica Mars* (2004–2007), for example, juxtaposes different styles of conversation among rich, working class, Latino and white, adult and teen, words are mainly used in our narrative to lead us to the themes that matter, they are not the theme themselves. In contemporary entertainment, only some of Steven Soderbergh's films such as *The Limey* (1999) and *Schizopolis* (1996), attempt to do what Milch does with language, but in a much simpler, shorter form.

There is a richness to Milch's language, always, a stylized voice not exactly of the setting of a series, but reflecting a romantic and deeper vision of the community that he explores. *NYPD Blue* took forms of New York speech with rhythms and pacing that seemed realistic in contrast to the dialogue of television of the day, but was stylized with a formality that might be compared to David Mamet, one of the few writers of the day who created a stylized naturalism recognizable to a general audience. Bureaucratic and professional language met up with working-class language, public and intimate language, with language of ethnic or cultural subgroup in ways that underscored characters and added sparks as different languages came together, but not as the subject matter that it becomes in *Deadwood.*

Deadwood covers much of the same territory as *NYPD Blue*, while seeming completely different; an exploration of people at the point where morality, ethics, and law are examined as members of a community are forced to interact with those who have the power to enforce definitions.

Peter DeVries supposedly once said that all of American history could be understood as a struggle between two sets of James brothers, Frank and Jesse, and William and Henry. (Historian Otto Pease might have said it as well. Possibly first.)

Deadwood would seem to put Milch on the side of the outlaw Missouri Jameses, but his roots are in institutions and community, how they are built, not how they are attacked. These are more the issues of the New England Jameses, especially William James's examination of human identity through a combination of psychology, religion, and philosophy, but also the literary dissections of individuals in culture by Henry. The interconnected nature of humanity espoused in William James's Pragmatist philosophy balanced by the literary consideration of individuals in the world of society is the core of Milch's creative vision.

Milch received a strong sense of the sweep of American literature, a love of poetry, and the epic sweep of history from Robert Penn Warren and the model of the disciplined life of a writer, reinforced by his other mentor at Yale, R.W.B. Lewis. Lewis was a scholar of American thought and the biographer of the Jameses, so it's not surprising that he fell hard for the Brothers James, especially the philosophical explorations of William.

Given his Jamesian orientation and the model of Robert Penn Warren's consideration of American history, it is possible to guess at the series-long arc Milch had in mind for *Deadwood.* Through the three seasons that aired, Al Swearengen, the apparent villain of *Deadwood*, does more to hold the community together than Bullock, the once and future force of social order. Future seasons likely would have shown the community develop and change through the rise of law and order that Bullock represents as the institutions of civilization replace the order of self-interest and "law of the jungle" that Swearengen manipulated to hold Deadwood together. This change would have played out as more an evolutionary inevitability, with strong elements of individual tragedy, rather than the triumph of order that one expects from television narrative.

Swearengen might well have become the television equivalent of his mentor's Willie Stark. While it's unclear what arc Milch had in store for him, it would seem that the next step toward joining the nation meant that the powerful in Deadwood were going to be testing themselves against the entrenched national powers like Hearst who decided to take what Swearengen had carefully husbanded along. The natural development of the series would be the journey of a big fish in a small pond into the bigger waters.

The hero who lives long enough to become the villain is a standard tale in American Westerns, but Swearengen is an antihero from the start who never quite becomes a hero and, one suspects, would become more sympathetic as he becomes more a tragic figure rendered irrelevant by the progress of nationhood.

Deadwood's revisionist history lesson was critically praised and popular, in both the U.S. and in international markets. It was not a success on the level of network television hit but it was a hit by the standards of cable, widely rebroadcast outside the United States and even more popular on DVD sets. Milch allowed himself to believe that he would end his life writing the expansive story of the United States, capitalism, and humanity through *Deadwood* for the rest of his life.

At which point HBO canceled the series.

Milch takes pains to remind us that the series was popular and profitable, especially overseas, and that HBO canceled the series for reasons that had nothing to do with the commercial or artistic success of the show. In interviews, the pain the cancellation caused Milch is clear, in part because of his feeling that he had not protected the people making the show but also because it seems clear that Milch saw *Deadwood* as his masterwork.

Even while canceling the show, HBO hurried to get Milch committed to do another series and it is hard not to imagine that Milch created *John from Cincinnati* to continue to explore the idea of the idea but also to teach HBO a lesson.

HBO wanted a surfer series. They felt that surfers could deliver the youth demographic. Milch says he hated surfers, the ocean, and beaches and expressed obvious scorn for the idea of making art to deliver teens to commercial exploitation.

Milch agreed to make the show.

The result is a fascinating failure of television but a very interesting DVD novel in a television form. Or, as someone raised by Minnesotans cannot avoid considering, *John from Cincinnati* might be one of the greatest acts of passive-aggressive creation ever committed in the televisual medium.

John from Cincinnati: A Novel of Ideas

John from Cincinnati (2007) was an attempt by Milch to more directly address the issue of the connection of the individual to the community through a religious story that seemed philosophical or psychological by turns. This was Milch's opportunity to explore the ideas he had considered since college when he read William James's *Varieties of Religious Experience*.

In his "Fifth Lecture on Pragmatism," William James suggests that the listener "suppose that the world's author put the case to you before creation, saying: 'I am going to make a world not certain to be saved, a world the perfection of which shall be conditional merely, the condition being that each several agent does its own 'level best.' I offer you the chance of taking part in such a world. Its safety, you see, is unwarranted. It is a real adventure, with real danger, yet it may win through. It is a social scheme of co-operative work genuinely to be done. Will you join the procession? Will you trust yourself and trust the other agents enough to face the risk?"[8] This is a favorite quotation that Milch has been known to repeat. In its original context, it was part of an argument against the philosophy of Monism, rejecting the idea that everything is God's will so everything is as it should be. James argued for a more pluralistic approach to religion and philosophy.

But it is hard not to see the quotation also standing as the argument for making a series like *John from Cincinnati* where the viewer—and the author—don't know where things are going. In this lecture, again in an original context of the nature of religious experience, James goes on to explain that the safe predictable comfort of monism is self-surrender, that the Pragmatist should embrace a riskier universe.

The author of *John from Cincinnati*'s televisual universe, Milch seems to have applied these philosophical ideas in to the aesthetic, seeing in himself the ideal James suggests we should aspire toward in that same Third Lecture: "It is then perfectly possible to accept sincerely a drastic kind of a universe from which the element of "seriousness" is not to be expelled.

Whoso does so is, it seems to me, a genuine pragmatist. He is willing to live on a scheme of uncertified possibilities which he trusts; willing to pay with his own person, if need be, for the realization of the ideals which he frames."[9]

James describes the universe and the author of that universe in a way very much like Milch himself and the series he makes, and the ideal characters in the series that he has used his power to make. He tells his audience the world is an adventurous place where a man risks it all for what he will create.

The appeal of that vision to a gambler like Milch is hard to ignore.

Milch explained to a university class that:

> I hate water, I hate sand, I hate swimming. But if I am to rest transparently from the spirit which gave me rise, I need to believe that starting from there I can get to the spirit, unless everything I've said to you is bullshit.
>
> So I went to work.
>
> And I said: "Listen, that surfing idea. I love that surfing idea. I love it. I just wanna . . . fool with it a little bit."
>
> They said, "Well, but it's young, it's hot!"
>
> I said, "And I love that. I love that. But . . . what if this was about . . . in the aftermath of 9/11, whatever God is, becomes aware that we have been so infantilized, we have been made such children by our addiction to media, which doesn't sink its roots deep, that we have become capable of a genocide."
>
> Now, they sleep through this shit anyway, the bosses. "That's great! That's great . . ." And they're looking around going "What the fuck did he just say?" "That is beautiful!" "But you'll have the surfing, right?"[10]

Milch was asked to do a surfer series to appeal to a young male demographic, and he perversely responds by creating a series that is as much about the commercially mediated world that would lead to that request as it is an inverted fulfillment of the assignment.

Milch discusses the ideal television show from the network perspective, even for HBO, as a show that attracts "the young who have the most disposable income and have the most free-floating yearning, desire to attach to one thing or another which can through certain stimuli be rewarded in mass entertainment."[11] In his lecture to the Religion and Media class, Milch says he has resolved himself to this apparent conflict between the needs of the artist and the needs of the networks, and he has realized that there is no more duality between the two than there is between the two teams competing at the Super Bowl; they may appear to be in opposition but they are both working together as part of the larger experience that is the Super Bowl. In the same way, artists have to realize that the commercial needs of the networks are part of the act of communicating today.

This doesn't mean Milch accepts the interference by network executives, or putting commercial interests over artistic ones, only that he has come to terms with them as part of the process. In writing *Deadwood*, Milch began his practice of accommodating one part of that process by refusing to turn over scripts to HBO until it was too late to make any changes.

Acceptance is not the same as surrender.

John from Cincinnati centers on a community built around surfing that is made up of the parents and grandparents of the target demographic. The "illusion agreed upon" is surfing. At the center of the story is a 14-year-old surfer savant who may be a Christ figure, but who is more concerned about trying to find a sponsor so that he can make money off his surfing by advertising to a young teen demographic. Shawn, the young surfer, is the center of attention for the rest of the people in the show, but the series is mainly older people living under the shadows of their own pasts.

Here we see Milch overtly giving HBO the teen lead it needs, but also narratizing the needs of the network in Shawn's need to attract an audience, and then building into that an adult show about ideas.

A divine figure and/or an idiot savant, John Monad is dropped into the banality of the everyday life of surfers, a group Milch has described as "the stupidest people on the face of the earth." The series begins with Mitch Yost, Shawn's grandfather, finishing a morning of surfing, only to discover that he has begun to levitate slightly off the ground. He's then accosted by the agent he blames for ruining his son Butchie's life, who is trying to get an edge in signing Mitch's grandson. Mundane miracles and the necessities of commerce pass side by side in a type of storytelling that comes closest to the conventions of magical realism.

Magical realism developed as a literary movement in South America juxtaposing the rationalism and Catholicism of Europeans with the richer folk religion and magical traditions of the indigenous culture. For some, magical realism is a subversive political art giving voice to the colonized by turning the language and symbols of the imperialists back against themselves in a clash between the descriptive and the metaphorical. For others, it is simply a reminder of the miraculous in the everyday. And for others, it is the origin of postmodernism.

In the United States, magical realism has been most visible in the writing of Jorge Luis Borges and Gabriel Garcia Márquez and in the occasional subtitled film like *Erendira* (1983) or *Like Water for Chocolate* (1992). Milch does not copy the form, but re-creates it as he juxtaposes the basic idea of the banal rationalism of contemporary culture—obsessions with money, fame, and sports—with an unnoticed divine presence dropped into the center of this world.

John might be compared to the patient in the 1986 Argentine film, *Man Facing Southeast*, who may be a divine figure or might be a lunatic, but who teaches and develops a following among the patients. (John might also be seen as Milch's mocking self-portrait as he attempts to bring great ideas to people obsessed with unimportant commodified entertainment.)

John From Cincinnati would make an interesting literary novel. It might have worked as a British series, with their tradition of social documentary shaping the way this fringe culture is perceived. But Milch explored ideas that weren't television ideas in a story that didn't clearly pay off in standard American television terms. While not as obviously disruptive a form as a musical cop show, it was a way of telling television for which audiences—and critics—had no point of reference except absurd sitcoms like *My Mother the Car* (1965–1966), *My Favorite Martian* (1963–1966), and *Alf* (1986–1990) where a supernatural figure of wisdom, apparently an idiot savant, is dropped into the everyday world.

In a series in which Milch's teaching came closest to preaching for some critics, even if they didn't understand the sermon, Milch depicts the need for community in humanity, and the need to connect through community with the divine, as a failing proposition because we do not hear and see what is right in front of us.

For those seeking the author's autobiography in the series, one of the central character in *John from Cincinnati* is Butchie Yost, the great artist of surfing who revolutionized the form before buying into his own hype and becoming a junkie. Butchie is befriended by John and, in his gradually awakening to a higher power, Butchie leaves the drugs behind and returns to his calling as a surfer. There's more than a bit of Milch's post-heroin story in that summary, and Milch's identification with Butchie is clear in his commentary tracks on the DVD set. Given that John advises Butchie and turns him around, the combined presence of Butchie and John may be read as Potter's Marlow and his Singing Detective in reverse. Where Potter saw the toxic dualism defeated when the stronger version of the self destroyed the other, Milch's belief that there is no duality, that both are the same is resolved as Butchie and John each help the other become more connected to the community.

The Meanings of *John from Cincinnati*

Language in *John from Cincinnati* is a stylized blend and clash of a border beach community, with surfer and Spanish blended with the language of advertisers, agents, drug dealers, therapy graduates, and dry-land transplants like retired cop Bill Jacks, who never get into the oceanic experience of the laid-back beach reality. The abundance of language is not primarily used to create character, although it does that; Milch overtly deconstructs language to consider how language is composed of symbols and how meaning is created in words and, through that, the way meaning is created in lives. Continuing from the symbolic meaning of gold in *Deadwood*, Milch takes the deconstruction deeper to explore the way we attach meaning to symbols largely based on past emotional experiences—one facet of Warren's "pain of the past making the transition to the future of joy." The series dwells on the way meaning is attached to symbols, in language, act, gesture, logos, and numbers.

At the simplest level, viewers watch John develop through the series like Jeff Bridges's alien in John Carpenter's *Starman* (1984). John is "an extraterrestrial," according to Milch, aware that the word applies to a divine creature as accurately as to an ET-type alien. John learns to speak by reusing the words that others have said to him. With a couple of exceptions near the end of the series, John only repeats back the words of others, often without grasping their literal meaning, and then develops a language of his own out of those found phrases that is not based on the literal meaning of the words but can be understood if one pays attention and remembers the context in which he first heard the phrases he uses. His words reflect the meaning that lay at the heart of the communication—"I care about you"—rather than the literal meaning of words—(a statement about bowel movements). This creates comic effects as it teaches the audience the difference between words and meanings. When Butchie gets tired of being told by John that "There are things I know and things I do not know," he tells John to say "'I don't know, Butchie,' instead."[12] For the rest of the series when asked a question by anyone that requires the answer "There are things I know and things I do not know." John says "I don't know Butchie instead."

Sometimes John seems to be having conversations of which we hear only hear his words, perhaps related to the Jesus of the Gospels who began sermons saying "Amen." (That is, he was essentially saying "I hear you, I get it," before he began to talk, suggesting for some that he was in conversation with God.)

John also develops a "stickman" symbol out of a zero and ones (binary symbolism of the digital internet era) that functions like the symbol of the cross in Christianity, while also apparently standing for the specific words of God, in one scene repeated to become the literal writing on the wall, that those who understand can read, while others see only ones and zeroes.

The idea that there's a single meaning to any story, let alone any symbol, is a central ideology of the American television narrative forms. Simple entertainment television is predictable and clear, the pleasure comes from being able to not think, to let the show wash over us. The tease of the double entendre is about as far as pleasurable ambiguity is allowed in traditional television narrative. By contrast, Milch turns to a literary model that forces questions upon viewers and—most important—does not offer answers. His symbols deliberately lack one-to-one correlations.

So viewers of *John From Cincinnati* argue about whom the different characters represent, starting with the title character who should be "Jesus" for many because he's the title character, but in fact acts much more like "John the Baptist." Or is he "Satan?" Or just a "wise fool" in over his head?

In descriptive terms, John (the name John Monad appears on his American Express card, in a wink and a nod to the William James lectures on monistic theology), first appears in the margin between United States and Mexico, between ocean and land, between night and day to tell Mitch Yost, former surfing great, that he should "get back in the game."

When John needs money or a credit card or a cell phone, it appears in his pants pocket. He tends to be a mirror for other people. He can make people forget things and make body implants get very hot. He surfs very well. People treat him as a child, as a kidnapper, as what they need him to be. John speaks a lot about his father's words, which no one will understand.

If we read the series as a Christ story, then it is confusing that John is the title character, because Shawn seems to fit the role of Jesus of Nazareth from apocryphal literature, starting from the moment he brings Zippy the bird back to life with his touch. Which would make John Monad John the Baptist to Shawn's Jesus.

You, reader, should feel free to go online and explore some of the blogged theories about the show, less voluminous than those of *Lost*, but sharing a similar obsession with disambiguation and the need to lock down one-to-one correlations. In doing so, it might be helpful to remember the experience of Harlan Ellison who was challenged at some public event because he had written a story that was a variation on the Greek myth of Prometheus. His accuser said Ellison was a plagiarist for stealing the story, only to then attack him for plagiarizing *wrongly* because Ellison had changed the character's punishment to one that was heart-related rather than liver-related.

Consumers of American popular narrative are not trained toward ambiguity and we don't always understand the idea of the artistic uses of allusion to earlier mythic or cultural forms and the creative need for an author to reinvent. The other characteristic of magical realism that U.S. media tends not to prepare viewers for is the idea that there can be multiple layers of meaning in a single symbol.

So it is comforting for us to learn that the character Locke on *Lost* (2004–2010) refers back to John Locke and a specific philosophy that the *Lost* character embodies—just as the title characters of cartoon masterpiece *Calvin & Hobbes* got their names from John Calvin and Thomas Hobbes, and stand for two philosophies—one of predestined salvation (here played out as the eternal innocence of the child, regardless of how homicidal his fantasies) and one that believed that the natural condition of man in the state of nature is "nasty, brutish, and short" (here a toy tiger who regularly attacks his playmate).

But *John from Cincinnati* asks viewers to consider that *in some ways* John is like John the Baptist but *in some ways* he is like the title character of *Man Facing Southeast*. In some ways, Shaun is like Jesus, but in some ways, he is like Lazarus raised from the dead, and in some ways, he is not a religious figure, he is a representation of a marketing construct, the artist as a brand. And Zippy the pet bird is Lazarus and the Holy Spirit and just a bird and Bill's dead wife, all at once.

As Milch explained to the Religion and Media class, "The key is that nobody understands—not in the show, not in the audience. That's the point. . . . *John*'s extremely annoying to some."[13]

Meaning Shaped by Past Attachments

Unlike *The Singing Detective* (1986) we don't see the specific experiences that shape these characters. We are led to understand the nature of past traumas by the effects we do see in the present, with occasional information to reinforce or contradict the assumptions we have made as viewers.

The casting decisions were intended to add to this frustration at yet another level, ideally leading to eventual greater enjoyment or enlightenment. Milch set the goal of casting actors recognizable for specific genre roles to illustrate one of the series continuing lessons; that our attachments of the past shape our reactions to new events. So Bill Jacks, the retired cop, is played by Al Bundy from *Married with Children* (1987–1997). Shaun's grandmother is played by the prostitute who bonded with Tom Cruise in *Risky Business* (1983), Rebecca De Mornay. Linc, the agent/promoter, is played by the kid from *Beverly Hills 90210* (1990–2000), Luke Perry. The Chemist is played by Dr. Johnny Fever from *WKRP in Cincinnati* (1978–1982), Howard Hesseman. Even the small part of an Hawai'ian drug dealer is Mr. Wu, the Chinese mob boss from *Deadwood*, Keone Young.

The issue of past attachments and symbols permeates much of what happens in the series. A homicidal drug dealer kills time in a car outside a hospital by listening to Andrew Lloyd Webber show tunes on the car stereo. This is a nice joke for those who remember that, before *Cats* and *Phantom of the Opera*, Webber and Tim Rice did a rock musical, *Jesus Christ, Superstar,* that also provided a modern reinterpretation of the Christ story. But the Webber music keeps playing—in a series with very little musical enhancement—and the car stereo's music becomes the impromptu soundtrack for the resurrected Shaun's escape from the paparazzi at the hospital. The music is either completely wrong or totally appropriate, depending upon how the viewer chooses to process the meaning of the music and the mood of the scene.

The key point here is that, even though *John from Cincinnati* fulfills the basic rhythms and structures of a television show it is far more interested in pursuing the themes more traditionally found in a philosophical novel.

Luck

After the unsurprising cancellation of *John from Cincinnati*, Milch began to develop a New York cop show set in the 1970s, with racial conflict, police assassinations, and the last days of open police corruption as the background. The show, however, was sidelined as he teamed up with Michael Mann to create *Luck* (2012), a now-canceled series about the world of racetracks and gambling starring Dustin Hoffman (with *The Singing Detective*'s Michael Gambon in a supporting, villainous role).

With *Luck*, Milch leaves behind the semiotic and philosophical musings of John from Cincinatti to return to a focus on the disparate and inadvertent community created from the outsiders and loners of the world of gambling. There is the obvious autobiographical background of Milch and his father being passionate horse owners and gamblers.

The most interesting element to consider, though, is the struggle for authorship between two very controlling creative voices. On the one hand is writer/producer David Milch, who has done everything possible to protect his personal vision in his previous shows, from withholding scripts from producers to overseeing the editing of his shows himself. On the other hand is director/producer Michael Mann, famous for his control of all aspects of his films. Milch explores the world through words and ideas. Mann reveals it in visuals and action. Both men have strong opinions and are used to shaping things to their own visions.

By the time of *Deadwood*, Milch had developed the sure defense against script interference of never turning in a script until shooting had begun. For Milch, in agreement with the views of Dennis Potter 30 years earlier, the superiority of television to film is that the writer is cut out of a film once the script is delivered and no film begins production until a script is in hand, at which point the director takes over. In *Deadwood* and *John from Cincinnati*, Milch could write and rewrite as a constant presence on the set, able to ensure his vision right through the editing room, where he would make his final script adjustments as he controlled the final cut of the show.

But Michael Mann is a film director who started out in television the central force behind *Miami Vice* (1984–1990) and *Crime Story* (1986–1988). He is the sort of film director who expects compete control of his narrative. Milch's script delay tactic did not work on *Luck*, where his withholding of the script was blamed for production delays and one episode being cut from the season, followed by an uneasy compromise where Milch would control the script up to its delivery, but then he was essentially banned from both the set and the editing booth, where Michael Mann's vision would be final. On that basis, the series is an interesting opportunity for auteur critics to ponder whether the writer's or the director's voice dominates in an equally matched contest. The series was canceled as the connections between the uniquely voiced individuals of the show's community were just beginning to become clear, with an episode that showcased Mann's characteristic use of music-supplemented action and montage to reveal character (even though he was not the actual director of the episode). It's impossible to know whether one vision would have ended up dominating the series or whether the two would have combined to provide something new from both men. As it is, later viewers who find *Luck* on its DVD set will be left to debate whether the series is a nine-hour Michael Mann film or a nine-chapter David Milch novel.

Chapter Eleven

David Simon and Fictional Nonfiction Television

In Which We Consider "Nonfiction Fiction" in an Era of Fictional News, the Greatest Show Ever on Television Not Being Television, and the Novelistic Expansions of *The Wire* and *Treme*

Nicola: *Write about real things in a realistic way—real people, real joys, real pains—Not these silly detective stories. Something more relevant.*
Marlow: (with contempt) *Solutions.*
Nicola: *What—?*
Marlow: *All solutions, and no clues. That's what the dumbheads want. That's the bloody novel—He said, she said, and descriptions of the sky—I'd rather it was the other way around. All clues. No solutions. That's the way things are. Plenty of clues. No solutions.*

—*The Singing Detective*[1]

The Wire is the greatest television program ever made.[2]

The Wire never won an Emmy.

That would be the simple proof that *The Wire* (2002–2008) was something other than a television show. Even as it became a cliché that *The Wire* was the greatest series ever to air on television, no less an authority than the Academy of Television Arts and Sciences seemed to think it was not television. In its five-year run, the "greatest television show ever" never received a single Emmy, and was only nominated twice, for outstanding writing for a drama series. (In 2005 it was defeated by *House* for the "Three Stories" episode and in 2008 it lost to an episode of *Mad Men*.)

One explanation is that this is the latest variation on the case of Stanley Kubrick; Kubrick was generally snubbed by the Hollywood establishment because he worked in England with non-Hollywood crews. *The Wire* was shot on the East Coast and used mainly East Coast and UK actors. Even its writers came from outside television, let alone Hollywood television.

Obviously, there are differences between great art and award-winning art—the classic example being the loss by Kubrick's *2001: A Space Odyssey* to Carol Reed's *Oliver!*, the musical, in the 1968 Best Picture Oscar race. But *2001* did get a nomination. There's something worth noting in who doesn't get invited to the party, who a culture doesn't recognize as one of its own. There's a theory that a biological imperative drives us to shun the thing that seems to be one of us but that we sense is somehow different.

The mixed reaction to David Simon's work includes high praise but not high television ratings, reflecting the uneasy positioning of Simon's sense of storytelling on weekly broadcast television. Over the years since *The Wire* began, it's been conventional wisdom that the main audience would find the series on DVD. After the second season of his next show, *Treme* (2010–), Simon made it clear that that is who the new series is written for. In an interview with Matt Zoller Seitz, Simon said he had to admit that he was "writing the show for people who have a complete season DVD set in front of them, or who are watching the show via HBO On Demand, or who can otherwise absorb it all as a piece, and watch [the episodes] all in a row."[3]

Simon's television career begins with *Homicide*, as the importance of the single episode on television was beginning to decline. Once the technology changed to allow viewers to watch multiple episodes in a row, writers have had increasing freedom to create shows to be watched that way. At some point, the weekly airing becomes only a transitional phase on the way to the actual object, the DVD set.

That is, the DVD novel.

Facts and Truth

Throughout his television career, David Simon has worked from two formal traditions; he has moved television writing toward the "episode as a chapter of a longer work" that he describes as a video novel. And he has built his shows from research like the journalist that he was. At a time when what we call journalism has become increasingly less factual, Simon represents an alternate trend, a style of fiction that aspires to serve nonfiction, a sort of "New-er Journalism" for the post-factual stage of American culture.

A journalist by training, Simon's television shows carry through on one of the main missions of good journalism, to educate and inform the audience through compelling stories about people and events that go overlooked in day-to-day life. After writing for the *Diamondback* newspaper at the University of Maryland, Simon moved on to 12 years of mainly crime reporting for *The Baltimore Sun*. Becoming increasingly cynical about the ability of daily journalism to effect any kind of meaningful change, Simon took a year off to follow the "murder police" of Baltimore's Homicide Division, writing the nonfiction book, *Homicide: A Year on the Killing Streets* (1991). The book is reportage that moves into the realm of ethnography, an immersive study of a community written to create the power and characters we expect of fiction, but without the artificial clarity of neatly plotted stories and resolutions.

Not surprisingly the book was received and reviewed as a nonfiction novel in the tradition of New Journalism such as Tom Wolfe's *The Right Stuff*.

One way of understanding Simon's writing process is to consider it as a move from reporting the facts to the perpetual challenge of journalism: how to reveal the truth, and, in Simon's case, to effect change. Presenting facts fictionalized in novels has been one avenue for attempting this—from the urban novels of Charles Dickens, to which *The Wire* is often compared, to the muckraking novels of Upton Sinclair and Frank Norris.

Taken in the other direction, the fictionalization from facts to express larger truths led to the scandals in journalism associated with a later *Diamondback* alumni, Jayson Blair, who became the poster boy for reporters making up their interviews and data when he went to work at the *New York Times*. The scandal of journalists lying to create the stories they wanted but couldn't find might be seen as a stage in the more general drift of American journalism toward the present state where cable-led "journalism" is as likely to present opinion and assertions about perceived truth as it is to present factual reporting.

Simon's approach provides a fascinating mirror image to the "fictions presented as fact" in contemporary journalism. His shows are created with an old-fashioned journalist's obsession with factuality and research, using that as the basis from which to blend and narratize the facts into fictions.

Simon makes clear that his shows are fiction, but insists on substantiating them by reference to specific factual points of departure. In interviews he and collaborator Ed Burns both regularly refer to the actual people or actual cases that inspire the stories, but he has also been adamant in asserting that, for example, Tommy Carcetti, the white Baltimore councilman who became mayor and then governor on *The Wire* was not a representation of Mike O'Malley, the present white governor of Maryland, who was elected after a term as Baltimore mayor, preceded by time as a councilman. Likewise, *The Wire*'s corrupt female council chair who wanted to be mayor herself, was not based on an actual female council chair who wanted to be mayor and was the subject of a federal corruption trial shortly after the series ended.

Where bad journalists gave in to the temptation to attach made-up quotes and fictional stories to actual people and events, Simon's work has reversed the process, using research and specific factual cases as the inspiration for fictional stories that reveal greater truths—but not necessarily facts—to viewers.

Robert Penn Warren's novel, *All the King's Men,* followed a similar path, taking his interest in American historical movements and his personal observations as he watched the Fascists come to power in Italy, and attaching those themes to a story about a Louisiana politician named Willie Stark, a character who bore quite a resemblance to the actual politician, Huey Long. (The historical Huey Long would have been a serious populist challenger to Franklin Roosevelt for the presidency if he hadn't been killed—by an assassin or accidentally by his bodyguards, depending upon your preferred version of the facts.) Warren repeatedly denied that his character Willie Stark was Huey Long, but most of what people think they know about Huey Long comes from Willie Stark, and I suspect that most of what people think they know about O'Malley comes from *The Wire*'s Carcetti.

It's an interesting side effect; the fictions can replace the facts. Which may be why Simon puts so much emphasis on grounding his fictions in research.

Homicide: Seven Years on the Screen

Barry Levinson (Baltimore-born director of Baltimorean films like *Diner, Avalon*, and *Tin Men*) decided to turn *Homicide: A Year on the Killing Streets* into a series for NBC. *Homicide: Life on the Streets* ran for seven seasons (1993–1999) with a television movie in 2000 that compressed a season's worth of plotting into two television hours. The show was critically acclaimed but never matched the ratings of *NYPD Blue* (1993–2005) on ABC.

A common explanation for the low ratings was that *Homicide* wasn't as exciting, but a more complete explanation is that, for all its advances, *NYPD Blue* gave viewers familiar television storyforms and television acting. By contrast, *Homicide* followed Simon's book in showing that homicide detective work was not what television had trained the audience to expect. Furthermore, the stories and acting styles did not do what regular television viewers expected and, for a medium intended to provide comfort in familiarity, *Homicide* was unsettling.

Simon was invited to write the pilot for the series but deferred and took a year learning television writing. He co-wrote his first script for the first episode of the second season with David Mills, a friend from the *Diamondback* who had become a Hollywood scriptwriter. Only in the fourth season did Simon quit his job at *The Baltimore Sun* and join the show.

While the series has a large ensemble cast of characters who all have their storylines and featured arcs, viewers enter the series with new detective Tim Bayliss (Kyle Secor), an inexperienced cop who has used his political connections to get a homicide job he doesn't deserve. Bayliss is partnered with the unit's best detective, brilliant interrogator Frank Pembleton (Andre Braugher), the antithesis of Bayliss, who starts Bayliss on his series-long journey to becoming his own man—professionally, but also in settling his personal issues.

By way of contrasting storytelling styles, *NYPD Blue* did several individual episodes that dramatically dealt with child sexual abuse, including one highly praised episode dealing with Detective Russell's revelation that she had been sexually abused as a child. By contrast, the final three seasons of *Homicide* followed Bayliss's dealing with the aftermath of his sexual abuse as a child, exploring its effects on his relationship with his family and his fellow police, his troubled adult relationships, his sexual troubles and confusions, and his turn to religion as consequences of that abuse.

The storyline that really established the difference between *Homicide* and other network television, though, was the Adena Watson child rape-murder investigation that became Bayliss's first case. Through most of the first season, Bayliss works it, identifies a possible killer, and then in an episode dedicated to the interrogation of their suspect, Bayliss and Pembleton are unable to break his story and have to let him go. The case is never solved and haunts Bayliss for the rest of the series, with Pembleton increasingly irritated by Bayliss's inability to accept failure and move on. In the fourth season, Adena Watson moves back into the foreground when a similar homicide occurs and then when the prime suspect for the Watson murder is discovered to have died, ending the possibility that he ever might confess or that the case can ever be closed.

The series developed fictional stories from the factual cases and characters in the book, guided by the reality that police work is a bureaucracy, that many murder cases are not closed, that most closed cases do not end in murder convictions, and that police are people with personal lives as important as their work. The series seemed at times to go out of its way to remove the pleasure the casual audience could take from it. Once Andre Braugher's Pembleton had emerged as the star of the series for his brilliant, commanding mental and verbal skills, his character had a stroke at the end of the fourth season that deprived him of all of his strengths for much of Season Five.

The reality of *Homicide* was a complicated, messy world that didn't fit neatly into television stories, and the series was shot to reinforce its difference from traditional network television. *Homicide* was shot to look "like reality." That is a constantly changing set of conventions, but the show stood out on television with a washed-out "video reality" look and handheld cameras. *Homicide* overlaid the *Hill Street Blues*'s neorealism with the visual language of what became the cinema vérité/New Wave style of the 1960s. The resulting unsensual imagery and jumpy editing made casual viewers more uncomfortable than the more saturated colors, and more familiar camerawork and editing seen, for example, on *NYPD Blue.*

Cinema vérité is exemplified by the film that set its rules, the 1960 *Chronicle of a Summer* by Edgar Morin and Jean Rouch, an anthropologist and a filmmaker, who rather than filming the ethnographic "other" turned the documentary camera on "us" and filmed the French in 1960. What makes it cinema vérité was the framing of the film. The director and the anthropologist begin the film on screen, discussing with their interviewer what they intend to accomplish. They return to the film at different points to ask questions and direct conversation, but then they return at the end of the film to show the participants in the film what they shot and then interview them about how accurate the participants thought the film was. The film ends with the two creators walking away discussing how it's just not possible to capture reality on film.

Homicide gave viewers a vaguely documentary experience of watching the story through a camera following the characters as they did their jobs. This feeling was made self-referential in the fifth season when a videographer, J. H. Brodie (Max Perlich), was added to the team. Brodie's presence explicitly draws viewers' attention to how a camera shapes reality. This theme comes to fruition in the fifth season episode "The Documentary," directed by actual documentarian Barbara Kopple, where Brodie showed the squad the documentary he had made about them—causing much amusement and discomfort. What could have been an excuse for a retrospective clip show instead exposes them by Brodie's fidelity to the truth they claim to live by. The episode ends with Brodie revealing he's sold the film to PBS, as the New Year's Eve murder rush phone calls begin.

More generally, the lasting effect of *Homicide* is seen in the influence of its casting. Shooting on location in Baltimore with most of the incidental characters coming from Baltimore, Washington DC, and New York, *Homicide* changed the way people on network television looked. An equivalent shift took place in film after Francis Ford Coppola's insisted on the casting of real New York Italians not "Hollywood" Italians for the supporting cast of *The Godfather*. For seven years, *Homicide* offered a much wider range of facial and physical types, especially of nonwhite actors, than Hollywood-based television had done.

The Wire

By the end of *Homicide*, Simon had concluded that "television does not allow you to tell stories." He pointed to the narrative limits of five commercial breaks an hour and network demands for victories and life-affirming moments.[4] Simon saw HBO—subscription television—as the best way to tell the stories he wanted to tell. He and Ed Burns adapted their nonfiction book, *The Corner*, into a miniseries for HBO. The show won three Emmys, which made HBO ready to support a series from the pair and convinced Simon and Burns to make a creative home at HBO.

In creating *The Wire*, Simon was willing to allow the series to be marketed as a cop show, but his goals were much greater. As he has said repeatedly, although he is often ignored, *The Wire* is about the failure of the institutions of urban society at the end of the 20th century.

Simon explicitly approached the series as a novel, enticing novelist George Pelecanos, and later novelists Richard Price and David Lehane, to join his goal of a visual novel.[5] They set out to write novelistically, with the story and characters gradually developing over chapters, with incidents setting up events that would pay off several chapters later, and an assumption that an intelligent audience would pay attention and remember. An audience watching in a novelistic way would accept a more complicated vision of the world than the feel-good endings expected on network television.

Simon points to the close examination of a "cold" murder scene by Bunk and McNulty early in the first season that occurs without explanatory asides for the audience. The reason for the scene only becomes clear near the end of the season, when the conclusions the detectives drew from the scene are revealed as they use the information to turn a key witness.

In one episode, McNulty explains to a somewhat disgusted Detective Kima Greggs how a partner can cover for the other's adulterous relationships with the "extraditions" excuse; half a season later, it comes back when

McNulty gets a call out of nowhere from Greggs's romantic partner asking about her extradition duty. In the third season, McNulty drops in on the supporting female police officer from the docks on the previous season with the assumption that the audience will remember the backstory.

Another example of nontraditional television writing is the moment in the tenth episode of Season Three when the story takes viewers into a gay bar. If the viewers look away for a moment, they miss seeing Deputy Commissioner of Operations William A. Rawls enjoying himself in the background. The detail was irrelevant to the episode. It fundamentally changes the viewers' understanding of the character, resonating backward and forward in the series, while at the same time being completely unimportant. And Rawls's apparent homosexuality is never mentioned again.

The Wire taught viewers that they actually had to watch the show, not just have it on in the background. And the show taught viewers that they had to decide what to make of the information they learned by doing so.

The Wire exploits the accessibility of the DVD set, knowing viewers have the ability to skip back or check older disks—or to go online—to find context for what they are seeing. In acknowledgment of the fundamental change in storytelling that was represented by *The Wire*, HBO agreed to send out the first five episodes of the series to critics before it aired, rather than the traditional one or two, so that reviewers could accurately reflect the uniqueness of the show.

Paying attention is one of the demands *The Wire* makes; the other is accepting a lack of clarity.

At the end of the third season, with Stringer Bell dead just as the police have finally caught him on the eponymous wiretap, McNulty and Bunk examine Bell's condo and can't imagine the man he thought he was chasing having anything in common with the man who lived there. McNulty realizes that he knew nothing about him.

"This is Stringer?" asks McNulty, "Who the fuck was I chasing?"[6]

This is a radical shift in television narrative. American television promises that even if we don't understand, our heroes do. And we will know eventually. That is the unspoken contract viewers have with American television.

The term for what *The Wire* offers has fallen out of usage; it's *aporia*. Beyond confusion, McNulty is left not just realizing that everything he knew about Stringer Bell was wrong, it's that he will never know what of what he knew was true.

The Wire's Literary Precedents

Where many critics have pointed to the literary precedent for *The Wire* in the writing of Charles Dickens, with his expanse of characters interwoven through extended stories of the modern city, there is another, more recent literary precedent for what David Simon does with *The Wire.*

There are 10 Swedish police procedural mysteries known collectively as *The Story of a Crime* officially, and less officially but more widely, as the Martin Beck novels. The novels were written by Per Wahloo and Maj Sjowall, a crime reporter and his common-law wife. Both were political activists who set out with the goal of writing 10 novels featuring detective Martin Beck, a detective dying of cancer, that reflected the social ills destroying Swedish society. Or, as Per Wahloo put it, to "use the crime novel as a scalpel cutting open the belly of the ideologically pauperized and morally debatable so-called welfare state of the bourgeois type."[7] Published, one a year, from 1965 to 1975, each novel is a consideration of one element of the failure of Swedish society, just as each season of *The Wire* would do.

Bleak, almost despairing, with a bare glimmer of hope across the whole series, these essentially humorless novels use one policeman's investigations and personal life to expose and comment upon Swedish society. The tone and the strategy of the novels ended up setting the conventions for Scandinavian detective fiction still obvious in recent international successes like the Swedish Stieg Larsson's trilogy of Lisbeth Salander novels, Henning Mankell's Inspector Wallander novels, and the Norwegian Jo Nesbo's Harry Hole procedurals. The influence is also seen in *The Killing*, the recent AMC remake of the hit Danish television series.

The Wire doesn't exhibit the Scandinavian moodiness, but it does share Sjowall's and Wahloo's basic fatalism about the collapse of social institutions explored through the largely ineffectual investigations of committed police detectives. Even more than his fellow journalists from Sweden, David Simon uses fictional television to do what journalism once did, at a time when American journalism more often seeks to replicate the cheapest of fictions.

Greek Tragedy on the Chesapeake Bay

Television drama derives from Shakespeare. At least, that argument could be made based on the general Harold Bloom-ian orthodoxy that Shakespeare created the idea of the individual. Earlier in the interview already quoted, Simon made the point that *The Sopranos* and *Deadwood* both draw from the Shakespearean sense of drama.

Others might want to adjust the argument, but there is that sense that in the 1500s some interesting cultural changes were taking place, including the Protestant revolution, that Shakespeare was reflecting. People were individuals, not extensions of their family or their class, their church, or their country. People could make individual decisions and would suffer the consequences of those individual decisions. Each individual charted his own path to Heaven or Hell. Macbeth decides to speed his ascension to the top and damns himself to Hell for the advancement of his colleague's son. Othello snubs Iago and ends up destroying his wife and his life. Hamlet can't decide and inspires generations of James Dean imitators.

The Wire's dramatic structure turns its back on that tradition and turns to an earlier tradition. David Simon explained: "The guys we were stealing from in *The Wire* are the Greeks. In our heads we're writing a Greek tragedy, but instead of the gods being petulant and jealous Olympians hurling lightning bolts down at our protagonists, it's the postmodern institutions that are the gods. And they are gods. And no one is bigger."[8]

The Wire depicts a world where individuals never really control their own destiny. Instead, we are the playthings of the failing institutions of contemporary America, all of which are collapsing and crushing us in their debris, without the comfort of traditional network feel-good moments. As Commissioner Ervin H. Burrell explains in the third season, "It's Baltimore gentlemen. The Gods will not save you."[9]

This single point may explain why, however much the series is praised, it rarely generates the likeability quotients of even mediocre network shows. At its core, American television tells us we can have anything we want if we only buy the right products and imitate the right celebrities. *The Wire* reminds us that we are powerless.

However beautifully prepared, *The Wire* is the broccoli in the television dessert buffet.

This is not Simon's version of David Milch's refutation of the individual, where the individual hero is replaced by the community. Simon denies heroes as television understands them and offers a journalistic sort of storytelling about people in context, people trying to get by in a world that doesn't much care if they live or die. Heroism is possible in this world, but the traditional hero is a falsehood.

McNulty nominally fills the role of the hero of the series, but his role in *The Wire* is to undermine viewers' expectations. To start with, McNulty is a screw-up who blames everyone else. While his self-destructive arrogance kicks the plot into motion in Season One, it also sabotages the case in the final season. Viewers are shown that McNulty is smart, but not nearly as smart as he thinks he is. And when Detective Greggs is shot, McNulty doesn't leap into control, he falls apart, and wallows in self-blame and self pity. As if to emphasize the point, McNulty is essentially left out of the second season until events are well underway.

None of the other characters become our heroic surrogate, either; there are too many and they all have strengths and weaknesses. This only makes sense in a series where, for all their efforts, most plans by cops and criminals succeed or fail based on incompetence and chance. The entire second season of the series, with an investigation into corruption among dockworkers with monumental consequences, only occurs because a petty senior police officer feels his donated stained-glass window has been consigned to the second best location in his church.

For many viewers, Omar is the hero, the gay killer who robs drug dealers and scares everyone, because he acts most like a Hollywood hero—skilled with violence, living by his personal code outside the system, doing what he wants, apparently unlimited by the power structures that hold back everyone else in the show.

Tellingly, this overlooks the character the show puts forward as its true hero: Bubbles, the junkie informant who Simon has admitted is a sentimental softening of reality. Bubbles's heroic journey is to stay alive, get straight, take responsibility for his actions and continue living, and reenter his family in the last episode ("to enter his house justified," as the John Ford Westerns would put it). While the character's heroic traits are always present, only in the final episodes of the final season as a *Baltimore Sun* reporter writes a feature story about Bubbles, does the series reveal, in retrospect, that he may have been the true hero of the series.

That's a novelistic consideration of heroism worthy of Robertson Davies's novel, *Fifth Business*. The novel that takes it's name from the term for very minor roles in opera, tells the story of a character who is barely a supporting player in the lives of important people. And yet he is the hero of his story nonetheless.

Urban Sociology as a Male Soap

The Wire explores the collapse of social institutions, but it also provides the opportunity to consider, for a final time in this context, the place of the male soap opera in longform television.

The female soap opera focused on the female suffering in the home and in the places of nurturing—kitchens, hospitals, coffee shops, and breakfast nooks—with focus on the betrayals of the heart and violations of marital/romantic commitments. *The Wire* completed the move to the male soap opera, albeit from a deeply Victorian definition of the genders. The world of masculine bureaucracy—the police, the union, the drug family, the newspaper—where, rather than love, the issues are honor and respect. The tactics may be similar, but the threat of physical violence and death hangs over the testosterone- rather than the estrogen-driven struggles.

This doesn't mean that there are no women; in fact there is a wide range of female characters in the series, including women much tougher than the men. Women who aren't "masculine"—tough or confrontational—end up marginalized. As Pembleton tried to teach Bayliss in *Homicide*, caring makes you vulnerable. The survived tragedy of the fourth season is that Prez, the cop who became a math teacher, cares about what happens to his students who are mainly beyond his reach. Prez persists and has an effect, but he remains vulnerable when last he is seen. The nurturing roles are more likely to be taken by men like Bubbles, whose caring almost destroys him, or the ex-con who becomes a boxing coach, with the limited hope of teaching some of the boys to reshape their rage.

In direct contrast to the man-boys of cable's attitudinal comedies, the men of *The Wire* are adult men. Even the boys are more adult than child. This world does not tolerate the weakness, innocence, or immaturity that make the man-boys of *Entourage* (2004–2011), *Californication* (2007–), and *Rescue Me* (2004–2011) so endearing to their audiences.

This shift to the soap opera of bureaucracy begins back in *Hill Street Blues* (1981–1987) but *The Wire* makes that battle within the male world its subject matter. The battle between good guys and bad guys is less relevant than the fact that they all fight from within institutions that confine them. The criminal mastermind Stringer Bell may become wealthy, but it is only to see it all stolen from him by bureaucrats and developers, lawyers, and politicos who he can't even begin to fight.

Whether it's politics or the police, drug dealing or the unions, there's no way to be what a man should be. It's generally noted that the series constantly juxtaposes criminal and cop, showing the essential similarity between them; what is often overlooked is the effect of this constant comparison. Over time, the differences between cop and criminal disappear and the shared, if inevitably doomed, struggle of individual men to succeed and maintain a sense of their own honor emerges as their common ground and inevitable tragedy.

The joke of Detective McNulty repeatedly being returned to harbor patrol is a nod to the network television reality that nothing ever changes, but the broader message of the series is that television reality gets only part of the reality of the modern city. Nothing changes, except to get worse.

Treme: Post-Devastation Storytelling

Treme looks at another contemporary city, New Orleans, after Hurricane Katrina. The first season starts three months after the hurricane. The second season takes place a year later, and each succeeding season is supposed to reflect the next year of recovery and rebuilding. The writers of the show have constructed fictional people living through the heavily researched social, political, and economic realities of New Orleans each year.

Treme finds the alternative to *The Wire*'s crime-centered story in the story of music in the lives of a community. The new series layers the stories, the characters, and the music to show viewers how music permeates life, in all the good and the bad. It's an atmospheric force that could not survive commercial interruptions and rewards immersive viewing of multiple hours. When critics compare Simon's shows to Dickens, it is with an emphasis on the number of characters in an urban setting, but *Treme* shares the same debt to 19th-century literature as John Irving's work does; a consciousness of characters existing in the passage of time.

The first season of *Treme* may be the greatest achievement of the DVD novel in that it uses all of the narrative possibilities of audio-video narrative to *show* us worlds but not *tell* us anything. Audio, video, dialogue, setting, and music reveal the story. The music establishes environment but also comments and provides plot, becoming metaphorically, and literally, the pulse of the show and the city.

It is also a series that requires and exploits the ability to tell a developing story over time. In response to the second season, some critics and viewers increasingly found the single episodes frustrating because they didn't do what television episodes are supposed to do. The focus on the longform story meant less concern with the closure of individual episodes; the characters and their stories did not offer the weekly clarity and feel-good triumphs viewers were trained to expect. Instead the series concerned itself with the small triumphs and tragedies of people, and how music buffers and translates the experiences. The dramatic force comes from the cumulative effect of seeing what changes and what stays the same over time; across episodes but also across seasons.

The Use of Time and Expectation in Creating Dread

There's another subtler element in the show's writing that works against traditional television writing. Network television has trained viewers to certain expectations. Even the "quality" television of the post-*Hill Street Blues* era reinforces clichés, albeit of the tragic nature of life, the two-sidedness of individuals, and the expectation of the reversal. A character makes a declaration that things are going to get better, and there will be an unexpected crisis. The monster is sweet to a child. A bad person defeated at the beginning will return once we've forgotten him to get his revenge at the end.

In the second episode of *Treme*, Albert Lambreaux (Clarke Peters), an old Mardi Gras Indian chief who is rebuilding a house, has his tools—his livelihood—stolen by a young punk. For students of film, the situation is the central crisis of De Sica's *Bicycle Thieves* (1948), perhaps the most famous of the Italian neorealist films where, in the middle of mass unemployment in postwar Rome, a father gets a job because he has a bicycle, but his bicycle is stolen and his life is apparently destroyed, though it must go on.

Neorealism is important to consider because it was, in part, a response against Hollywood films that told unreal stories of unreal crises, where neorealism told everyday dramas about everyday people. From the distance of over half a century, the films are as melodramatic as the Hollywood films they responded against, but they created a new sort of melodrama that eventually changed the sorts of stories that American film and television told in the 1970s and after. Neorealism used the audience's expectations of the conventions of Hollywood—simple moralities, recognizable character types, plot conventions, and Hollywood endings—against the audience, shocking them with a new set of more realistic conventions and moralities where, among other things, good people suffered, justice didn't triumph, and perseverance was about as much of a happy ending as you could expect.

On *Treme*, Lambreaux is able to get the tools back from the man who bought them from the thief. But Lambreaux then tracks down the thief, named Skinny, in an abandoned house where Skinny is trying to steal the copper wiring. In the empty dark house, in an act of extended violence, the old man uses a pipe to beat the young man to the ground and leaves him unconscious.

Traditional television has trained viewers to expect Skinny to come back after we have forgotten him, most likely to wreak some escalated revenge against Lambreaux. (An alternate, less likely, convention is that he comes back as a threat only to reveal he has turned his life around and he thanks Lambreaux for making him see the error of his ways.)

Treme never acknowledges this convention, but for the rest of the first season, Lambreaux lives alone in an empty dark bar, shot in a way that echoes the empty dark house in which Lambreaux beat down Skinny, and reminds viewers of the old man's vulnerability. We wait for a retaliation that never comes, which keeps us from enjoying his small successes as they only suggest another convention of "quality television"—that when things start to get better, something horrible will happen.

Treme's writing reworks the conventions in unpredictable ways. Telling the story of the aftermath of Katrina in New Orleans is a broadly depressing story with moments of optimism and sporadic happiness. A central arc of the first season is the search by a family and their lawyer to discover what happened to a young man who was arrested just before Katrina hit and has not been seen since.

Network television has trained viewers to expect a miraculous happy ending after much anxiety; "quality television" has trained us to know that the young man is dead. But the show keeps not giving viewers either ending. Instead, through the arc of the entire season we watch people gradually getting closer to what happened to him. The result is that we are continually brought to the point where a traditional "quality television" drama would reveal he's dead, but *Treme* doesn't do that. Eventually, the viewer is forced to let go of the trained response and think that may be the twist is not that they'll tell us he was dead all the time—the "quality television" cliché—but that they'll discover he's still alive and the family can be reunited in the network television cliché.

The result is that the viewer starts hoping while not believing, forcing the viewer into the same emotional state that many of the characters in the show are in. Without giving away other plot points, several of the multi-episode storylines, and a few of the single-episode arcs are structured in a way that leads viewers down a familiarly clichéd television path and then suddenly reminds us that that's not what we're watching after all.

The name for the slowly growing feeling the show creates is "dread," but it's a dread that is blended with unjustified optimism and the recognition that things and people carry on.

The only similar use of this strategy of dread created by playing off television conventions is a simple nontraditional choice made in writing the British series *Spooks* (*MI-5* in the United States, 2002–2011). The choice made for *Spooks* is that anyone, even a lead character, may be killed at any time in the season or in an episode. Only one original character makes it to the end of the series; every other character dies or must abruptly disappear to avoid prison, generally when the audience genuinely least expects it. The result is that a viewer has no way of knowing whether any character will make it to the end of episode, let alone the season. The anxiety this simple decision can produce for viewers trained by television to expect significant deaths only at the end of a season or a dramatic arc is wonderful to behold. It also reinforces the series message that intelligence work destroys everyone who gets involved in it.

The similar use of dread in *Treme* also, formally, adds to the importance of music because music becomes a relief—a safe zone from anxiety—for viewers that reinforces the use of music by those trying to rebuild their lives in the show.

The series teaches viewers to distrust clear explanations. Characters who explain what is happening in New Orleans might be right, but they are contradicted by others who seem to be equally correct. Even worse, the most articulate explainer of what has happened and the voice of one facet of the city's anger, Creighton Bernette (John Goodman), an English professor who becomes a viral Internet sensation for his drunken rants, kills himself without warning, leaving his wife and daughter behind with no explanation.

This event near the end of the first season of *Treme* is worth considering in comparison to the much more controversial death near the end of the first season of *Game of Thrones* (2011–). In both cases, a likable character was killed. The difference is that the first season of *Treme*, and *The Wire* before it, had prepared the Simon audience for unhappiness and unexplained tragedy. While *Game of Thrones* also rejects simple happy endings, the history of knights-and-dragons popular narratives and an audience drawn to the new series as typical television escapism, led to a sense of betrayal by viewers that was absent from responses to Creighton's death.

In the final episode of the first season of the *Game of Thrones*, Eddard Stark (Sean Bean) is clearly going to be executed, so viewers counterintuitively know that he will be saved at the last minute. Television has trained viewers to believe in the happy ending. The twist ending is that there is no twist. That is the shock. The lovable character is killed.

This working against the audience's training is the opposite of the subtler way that viewers discover Creighton's suicide. On the way to his suicide, Creighton spends too long on screen alone, and apparently aimless, for viewers to be comfortable. He doesn't do any of the obvious pre-suicide actions or give off the clichéd warning signs viewers would expect on a television show. Instead the form provides a clue: he is alone too long. More important is Creighton's small size in the shots; he is smaller in the shots than normal, which reinforces the idea that something is off, but not what.

Again, this is the creation of dread. Viewers are left to ask themselves what's wrong; did they miss a clue? If viewers get as far as thinking he is about to kill himself, they are forced to try to find a reason why Creighton would do that. The shots are held for a long time and there are not the interruptions of commercial breaks, so viewers know they will not get relief. They are stuck with this story.

After Creighton has killed himself, the effect continues. Viewers don't know why he killed himself and are forced to consider what they missed, which might lead some to a moment of revelation. Others might find a sick joke. Because one explanation would be that viewers are asking the wrong question. Given the inevitable despair in New Orleans, perhaps the question should be why didn't he kill himself sooner?

Creighton's specific situation is quite different from the other characters and yet something of it echoes with that of Albert Lambreaux's attempts to make sense of things as he attempts to pull his community back together, or Janette Desautel (Kim Dickens), the chef who cannot keep the bills paid even as she receives raves from the culinary elite, or Antoine Batiste (Wendell Pierce), the trombone player who can barely make a living in New Orleans.

Treme does not bring the characters together, but it shows a range in juxtaposition that causes the viewer to ponder the connections and the larger reality. This is the sort of narrative strategy that uses the DVD set to its best advantage, allowing the episodes to be read like chapters in a novel, with all the continuities and overlaps that disappear in weekly viewing.

Chapter Twelve

Endings

In Which Closure is Attempted, Film is Finally Addressed, and Speculation Abounds

I'm fascinated with the speed with which something like television feels like it's going the way of the dodo.

—Neil Gaiman[1]

With luck, the previous chapters have charted out the rough path that American television has taken over the last three decades as it moved from conventional episodic television into what has become a new model for what airs on television, but is no longer "television;" a type of longform narrative that, for lack of a better alternative, I'm calling the DVD novel.

It may be that the true DVD novel will never come into independent existence; that is, that we never get a direct-to-DVD 12-hour longform narrative. The economics would make that a safe bet because, so far, television has been necessary to pay for the production of longform televisual narrative. Further, the smart money in new media would argue that all digital media will soon exist in clouds on the Internet, and the physical object of the DVD will quickly be passé.

I may be too locked into a 20th century mindset, but I do think Americans like to own physical things that they can hold and look at on a shelf, so I expect DVD sets will last longer than techno-futurists might predict. If the DVD novel does become the digital novel, watched on the same iPad or Kindle screen that we use to read the latest "word novel," the distinction between the print novel and the televisual novel will have even further collapsed.

The point is that the existence of the DVD set has led to the inescapable reality that it, not television broadcast, is the final state of a television series and that has changed the way that television series are made and watched. Obviously longform television existed before the DVD but the boxed set made it easier to appreciate the benefits of the new type of television narrative, encouraging viewers to stop watching weekly broadcast television and to think of television shows as a longform story that is kept on your shelf and watched when and how they feel like it.

Single-episode television is still the dominant goal of network television, but even those shows have mainly incorporated at least one long-arc story and embraced the narrative by-products of longform television, like dramedy and unlikable heroes.

Film as DVD Novel

The other side of the DVD novel that this book has largely ignored is that film has also been changed by the DVD. When the final state of a film becomes a DVD sitting next to a DVD set of a television series, the distinction between the two increasingly disappears.

Thinking of the final form of a movie as a DVD—or part of a DVD set—makes it that much easier to make film series, or even break a single story into two films, knowing that they will end up being reunited on the DVD shelf.

We live in a culture with a taste for longform narrative. Perhaps a generation raised on Harry Potter novels is to blame. Video games also succeed by offering—on that same home screen—an immersion in a longform narrative, albeit interactive, that is built on an episodic or "level" structure but tells a longform story requiring hours of all-day or all-night experience that threatens jobs and relationships.

Regardless, from *The Lord of the Rings* to the *Bourne* trilogy, audiences have developed a taste for longform stories across multiple films made even longer when they are transferred into DVD sets. Obviously *Star Wars, Indiana Jones*, and *Back to the Future* are film series that began before DVDs but quickly took advantage of the boxed-set implications. Since the normalizing of the DVD set, we have also seen the *Twilight* franchise, the *Hunger Games* adaptations and, most influential of all, the *Harry Potter* film franchise. All of these series habituate viewers to expect long and continuing televisual stories with novelistic dramatic arcs.

This longform logic continues in the DVD releases of recent Marvel comics films, including *Iron Man 2* (2010), *Thor* (2011), and *The Avengers* (2011). The DVD releases include short interstitial films about a marginal SHIELD agent who appears in all of the feature films and is used to connect them all into one extended narrative. Only those who buy the DVD get the full story.

Taking the idea of films as raw material for DVD sets even further, the final novel of the Harry Potter series was split into two films, *Harry Potter and the Deathly Hallows: Part 1* (2010) and *Part 2* (2011), to be later reassembled on the DVD shelf for the novelistic closure that the viewers expect. Peter Jackson followed by splitting *The Hobbit* into two films in 2012 and 2013, as did the adapters of the final *Twilight* novel, with films in 2011 and 2012. DVDs have made it possible to double the length for the cinematic adaptation of novels.

Quentin Tarrantino's *Kill Bill* (2003/2004) was just over four hours long; his producer convinced him to cut it into two films. Jean-Francois Richet made two films of about two hours each about the famous French gangster Jacques Mesrine. The films, known in the United States as *Mesrine: Killer Instinct* and *Mesrine: Public Enemy No. 1,* were released one month apart in 2008. At the same time, Steven Soderbergh made *Che*, a two-part film about two different periods of Che Guevara's life that totaled nearly four-and-a-half hours. Both parts were released in 2008, with joint and separate screenings. It is only now that these films are on DVD that most viewers can watch them as the unit they were intended to be.

Films lengths have long been dictated by the need to fit the screening schedules of theaters; anything over two-and-a-half hours means one fewer screening each day, which means a significant loss of profits. DVDs can be as long as they need to be; in fact, making a film longer on DVD by restoring deleted scenes is seen as adding value to the DVD. The interminable ending(s) of the third *Lord of the Rings* film becomes an appropriate ending after viewing the 11-and-a-half hours of the extended DVD versions of all three films.

This practice of adding scenes for DVD release means that the true complete version of such films is never shown in the theaters, it is the one that appears on the DVD. Seeing a movie in the theater becomes the act of paying for a preview of the complete film that is eventually coming to DVD.

Ridley Scott, who makes both film and television through his Scott Free Productions, took it a step further when he released *Kingdom of Heaven* (2005) to the theaters with an hour missing. The complete three-and-a-quarter hour film he made could only be seen on DVD. His *Robin Hood* (2010) was only available in the unrated—and uncensored—extended version on DVD.

A key turning point in the primacy of the DVD over the theatrical film may have occurred with Zach Snyder's adaptation of Alan Moore's graphic novel, *Watchmen* (2009). Before the film was released, it was announced that the DVD version of the film would be significantly longer than that shown in theaters. This appears to be the first time a film was promoted by telling potential customers it wouldn't be as good as the DVD set.

In fact, after a DVD release of the theatrical version of the film (163 minutes), a director's cut was released (186 minutes). Then, about a year later, the ultimate edition of the film appeared as DVD set, adding still another half hour by finally including the gruesome parallel comic book story of misguided revenge that one of the characters in the film is reading. The juxtaposition of this material with the main plot functions roughly like the detective novel in *The Singing Detective*. It is not central to the surface plot but, when it is removed, a complicated self-referential commentary within the story disappears and it becomes more "story" than "literature."

David Lynch took another tack with *Inland Empire* (2006). Knowing his three-hour surrealist film would mainly be watched on DVD, he shot it to look best on a television screen rather than a movie screen and released the "too long, too confusing" film in theaters mainly to gather reviews and publicity for the DVD release. From the start, Lynch encouraged his audience to watch the film the way they would read a novel, using the pause button to set it aside from time to time in a way that cannot be done in a movie theater screening.

DVD sets have helped break down the distinction between film and television in another way. *Carlos* (2010), the international hit film about a terrorist, was originally released as a short television series, a pair of films, and a single film. Different DVD sets, from three to five hours, allow viewers to choose the narrative medium and length of story they prefer. Most Americans know *Das Boot* (1981) as Wolfgang Petersen's two-and-a-half hour submarine film but it is also available as a three-and-a-half-hour DVD set, or in its original form as a five-hour German television series.

Finally, film director Todd Haynes's five-hour adaptation of the novel *Mildred Pierce* (2011) for HBO may have been on television but was billed and reviewed as "a Todd Haynes film," which will sit comfortably on the DVD set shelf between the similar length *The Hobbit*, and final *Harry Potter* and *Twilight* double films.

Novels into Longform Television

If *The Singing Detective* was the flashy British explosion that stunned television writers and audiences with original novelistic television, it was the British adaptations of novels that *Masterpiece Theatre* steadily aired on American public television that more subtly converted viewers to novel-based television.

Television series adapted from a series of novels have been a natural transition, perhaps because most serial novel characters change as little as the characters of traditional network television shows. Most such adaptations use the novels as raw material, rather than actually adapting the personal and novelistic specifics of the books.

Three more respectful adaptations, *True Blood* (2008–, based on Charlaine Harris' novels), *Vampire Diaries* (2009–, based on L.J. Smith's novels), and *Dexter* (2006–, based on Jeff Lindsay's novels) have still made significant changes from the novels. While the first season of *Dexter* stays fairly close to Lindsay's first novel, they have since diverged so completely that Lindsay regularly has to defend himself from complaints about the television series killing off characters who are still alive in his novels.

The commercial success of these series have demonstrated the advantages of using a popular novel series as a starting point for a television series. The author has solved any problems of continuing the characters and situations across several novels meaning that there are already several seasons' worth of potential storylines that demonstrate where the series can go. Such shows also inherit a committed audience of fans.

At the end of 2011, at least a dozen novel-based series were being developed into network television series, the most visible being an NBC series based on Thomas Harris's Hannibal Lecter novels. It is worth remembering that many more shows go into development than ever come out again, but there is a greater openness to using novels for network television, even if only as raw material. In contrast to the British model of making classic novels, American television is turning to contemporary fiction. When classic fiction is made into television, it will most likely appear in the style of ABC's *Revenge* (2011–), the modernized female-centered remake of Alexandre Dumas's *The Count of Monte Cristo*.

This should not be seen as a turn to using novelists to write television, though. The series are being developed from novels through the standard group writing process with little or no involvement with the novelists. Some of the series, like *Dexter* and *True Blood*, have novelistic season-long arcs with closure, but most of these series will be episode-based with the goal of running forever without closure.

There are also novels being more closely adapted into short series with closure on the pay cable channels that will follow the success of HBO's adaptation James M. Cain's *Mildred Pierce*. Carl Hiaasen's *Skinny Dip* and Jonathan Franzen's *The Corrections* have been announced. At Showtime, Stephen King's *Under the Dome* is in production.

The major developments in novel adaptations are seen in the work of name brand novelists—Stephen King, George R. R. Martin, and Neil Gaiman.

King has long been open to experimenting with different types of transmedial delivery of his stories. Besides seeing most of his novels adapted into films and television miniseries or both, in the case of *The Shining* (1980 and 1997), he has also written and directed films. In 1996, King experimented with the Dickensian serial novel by writing *The Green Mile* as a series of monthly paperback "chapters" before republishing it under a single cover, and then seeing it turned into a Frank Darabondt film.

King has his own approach to the "wide-ness" seen in *Lost* and *Heroes*. His short story "N." was adapted into a serial narrative of 25 short limited-animation segments for cell phone viewing, later moved online, then packaged as a single DVD supplement to his 2008 short story collection, *Just after Sunset*, which included the print version of the story. Finally a stand-alone graphic novel of "N." was released, based on the limited animations.

Having already seen his 1978 postapocalyptic novel, *The Stand,* rereleased in a "complete and uncut edition" in 1990, adapted into a television miniseries in 1994, adapted by Roberto Aguirre-Sacas into a series of comics beginning in 2008, it was announced in 2011 that the novel will be adapted into three feature films directed by Ben Affleck. Which will no doubt end up as an extended edition DVD set.

The most significant King adaptation news of 2011 may have been the on-again-off-again on-again project to adapt King's *The Dark Tower* into three feature films and two HBO television series, all of which will come together as a set of DVDs for home viewing. If the project is completed, it will largely eliminate the distinction between film and television on DVD.

Neil Gaiman first came to prominence for his writing of the 75-issue comic, 10-volume graphic novel, *The Sandman* (1989–1996), which continues to be regularly discussed as a natural subject for television adaptation. Since *The Sandman*, Gaiman has written novels, films, and television, including his own miniseries *Neverwhere* (1996) and episodes of *Deep Space Nine* and *Doctor Who*.

In the summer of 2011, it was announced that Tom Hanks's production company would be producing a six-season HBO series based Neil Gaiman's 2001 novel *American Gods*. The novel is based on the idea that gods came to America with their believers and, as people lose faith, the gods lose power.

Just as with George R. R. Martin's deal for the *Game of Thrones* series, Neil Gaiman, will participate as a writer and executive producer. Both men wrote their epic novels after having experiences making films that led them each to turn down repeated attempts to turn the novels into films, knowing that the novels would not fit into a film.[2] Both saw premium cable television, essentially HBO, as the only place their novels stood a chance of decent adaptation. Their involvement does not give them the power to overrule production decisions—as Martin points out, he could not stop them if HBO decided that next season *Game of Thrones* would have aliens come down or everyone would turn into vampires, but he trusts the producers to generally respect the source material.[3]

Gaiman and Martin, like the younger comic- and television-writing playwright Roberto Aguirre-Sacas, would seem to be ideally suited to exploit the possibilities of a true DVD novel. Each has successfully created narratives in all of the related media. Martin and Gaiman also have enough fame and a large enough fan base that they could potentially create a profitable direct-to-DVD if they were interested, just as Louis CK proved that a comedian could make over a million dollars by directly selling his online concert to fans for a five-dollar fee.

The other precedent for thinking this way is the success of Joss Whedon's *Dr. Horrible's Sing-Along Blog* (2008), an approximately 45-minute musical about an aspiring supervillain who is denied his true love by a loutish superhero. The series was created by Whedon and friends during the 2007–2008 Writers' Guild Strike. Its three acts were available for free online, then sold through iTunes, and finally released on DVD. Whedon's fan base and that of actors Neil Patrick, Felicia Day, and Nathan Fillon and the novelty of the project generated enough publicity that the project made a profit.

The Internet is clearly the location of interesting developments, although televisual narrative faces the same difficulties as online graphic novels and online music; how do you attract an audience and how do you make money?

So far short-episode comedy series have been most successful. Two short-episode comedy series that have done well at creating cumulatively longform narratives are the gaming-themed series *The Guild* (2007–) and *Legend of Neil* (2008–2010), although the series that have crossed over to cable television—*Web Therapy*, *Secret Girlfriend*, and *Childrens Hospital* have been built on freestanding episodes.

Heroes, *Lost*, *The Walking Dead* (2010–), and *Battlestar: Galactica* (2004–2009) are among the television shows that successfully produced dramatic short-episode web series, but they benefited from the publicity related to their television shows. Keifer Sutherland and John Hurt had the star power to attract publicity and an audience to Hulu for *The Confession* (2011), a dramatic web series of 10 five- to nine-minute episodes. *Aim High* (2011–) has aligned itself with Facebook and created a socially interactive web series that integrates the "wide-form" model of *Lost* and *Heroes* into the series itself.

The continual deterioration of the separation between television production and television distribution in relation to the Internet, seen most clearly in the purchase of NBC by Comcast, creates interesting new dynamics for the creation of longform television on the Internet. So far, the networks have been airing online episodes to test series for possible transfer to traditional television broadcast. Such once unlikely entities as Netflix, Amazon, and Google/YouTube are beginning to explore production and direct distribution of narrative fiction. YouTube has been attempting to attract unpaid films and series to premium channels. Amazon is offering Amazon visibility and prizes to attract free content.

But Netflix has come closest to recreating HBO's original cable breakthrough on the Internet as it has begun to produce its own original programming to feature on its streaming video site. Well over half of Netflix's streaming video viewers are already using the service to watch seasons of television series at their own pace, but their production of original programming for the streaming audience has moved them ahead of any Internet competition. Netflix's production of the reboot of *Arrested Development* (2003–2006, 2012–)received the most media attention, but their series *Lilyhammer* (2011–) with Steven van Zandt playing a variation on his *Sopranos* character as a mobster in Scandinavian witness protection has also been drawing attention. Netflix revealed a change in longform television power when it outbid HBO for a 26-episode David Fincher and Kevin Spacey–led adaptation of Michael Dobb's novel, *House of Cards* (2012–2013) which was previously adapted into a four-episode series by the BBC (1990). Still, Netflix began 2012 exploring the possibility of creating it's own cable television channel to directly compete with HBO on its own turf.

In practical terms, television is being absorbed by the Internet. Both come into our homes through the same "intertubes," only one of them, television, attempts to hold on to the time-limited rigidity of a broadcast schedule.

The DVD Novel and the Legacy of *The Singing Detective*

In 2011, novelist Salman Rushdie announced he was writing *The Next People*, a science fiction series for Showtime. Citing *The Wire, The Sopranos,* and *Mad Men* he pointed out that, unlike films where funding is mainly available for "comic book movies," serious adult fiction could still be created on American television. He explained that "Matthew Weiner in *Mad Men* writes the entire series before they start shooting, and if you have that, then what you can do with character and story is not at all unlike what you can do in a novel."[4]

Awake is an NBC series that premiered midseason in 2012, that seems to provide a sort of closure for this book. In *Awake*, a police detective suffers a trauma and ends up living in two different realities; one life in which his wife is dead and his son is still alive, and another life where his son is dead but his wife is still alive. He works different cases—and sees different psychiatrists—in each of these realities but they gradually start to bleed into each other.

The hook is that one of the realities is apparently a dream state he has created to deal with the trauma of losing a loved one, but he doesn't know which one and he doesn't want to lose either because this way he is able to be with both of the people he loves.

The series has the potential to be the next *Lost*—a mystery based on shifting reality, with police cases to be solved within the episode and the long-arc story of an apparent conspiracy that caused the death of his loved one/s, and a series long arc to determine which is the "real" reality. So far it is a strong show with many different directions in which it could go.

But what strikes me about the series is that *Awake* seems to be what you would end up with if you pitched the idea of *The Singing Detective* to several levels of very talented American television writers and network executives. They'd get rid of the bitterness and the disease, substitute a car accident. They'd get rid of the exploration of the creation of identity through childhood trauma and the inescapability of the past by focusing instead on a recent trauma that happened to an attractive adult who's looking toward the future. They get rid of the powerless whiny invalid writer who fantasizes about being a powerful detective and you replace him with an actual young virile powerful detective, able to solve cases even while traumatized. The hallucinations are an escape instead of a torment and the healing is not an act of isolated struggle, it is an act of understandable love for wife and child.

Rather than seven intense and disturbing hours, the series is designed to run forever with twists and turns to keep the audience hooked while still providing episodic closure at the end of each hour.

I opened this book suggesting that television is two things, it is a thing, an object, and it is the content that is distributed through that object.

In the chapters since then, I've alluded to another thing that television is: television is a mindset. American television tells us that everything is okay, that we are right, that the world revolves around people like us, that things work out for good people, that bad people are punished and tragedies have explanations. Television tells us not to think but to enjoy, and everything on American television should be "family entertainment," suitable for everyone.

There is nothing inherently wrong with this mindset, but it seriously limits the sorts of stories that can be told. Longform television can be made that continues to reinforce the values of this mindset, but it also offers the ability to tell stories that show multiple perspectives, to juxtapose incidents over time, to raises the possibility of challenges to that mindset.

In the final analysis, the DVD novel might need to exist to give permission to expand the range of television narrative, to challenge the normal way of telling stories, and the normal values that network television has trained us to expect. It may be that televisual narrative that doesn't "act like television" needs to be physically someplace else to do that.

As Dan Harmon, the creator of *Community*, explained:

> I grew up watching TV. Saying something is TV is like saying it's a sonnet or a haiku. It's got its rules, and those rules, when obeyed, are part of what makes the thing beautiful. You use it to communicate to someone. If you say, "This is a sonnet I wrote for you," then you better not give them a limerick. If you say, "This is a limerick that I wrote for you," you're using that medium.[5]

The DVD novel is making it easier for some people to stop writing limericks and try sonnets, instead.

Notes

Introduction

1. Matt Zoller Seitz, "Hot Seat: David Simon Explains 'Treme,'" *Salon*, July 4, 2011, www.salon.com/2011/07/04/treme_season_2_david_simon_interview/.

Chapter 1

1. Bruce Bauman, "The Critic in Winter," *Salon*, January 2, 2003, www.salon.com/2003/01/02/fiedler.
2. David A. Perdue, "The Serial Novel," David A. Perdue's Charles Dickens Page, November 2, 2011, charlesdickenspage.com/works.html.
3. *The Promise: The Making of Darkness at the Edge of Town*, directed by Thom Zimny (Los Angeles, CA: Columbia, 2010), DVD.

Chapter 2

1. The Museum of Television & Radio Seminar Series, *The Television of Dennis Potter: Seminar No. 2 of 4: Potter and Bochco: A Cross-Cultural Dialogue* (New York: Paley Center for Media, January 15, 1992).
2. *The Singing Detective*, directed by Jon Amiel (BBC Warner, 2003), Commentary track, DVD.
3. Clark Collis, "Prime Suspect," *Entertainment Weekly Fall Television Preview*, (September 9, 2011), 92.
4. *The Singing Detective*, directed by Jon Amiel (BBC Warner, 2003), Commentary track, DVD.
5. Dennis Potter, *The Singing Detective* (New York: Vintage, 1988), p. 86.

Chapter 3

1. Steve McKerrow, "'Edge' Is Honest about TV Industry," *Baltimore Sun*, February 12, 1992, http://articles.baltimoresun.com/1992-02-12/features/1992043261_1_steven-bochco-dennis-potter-singing-detective.

2. William Grimes, "A Feast of Wit? Not Quite. Bile Flows," *New York Times*, January 18, 1992, sec. 1, p. 13, col. 3.
3. Pamela Douglas, *Writing the TV Drama Series: How to Succeed as a Professional Writer in TV* (Studio City, CA: Michael Wise Productions, 2007), 63.
4. Freddie Prince, "Washington Heights," *Looking Good* (New York: Columbia Records, 1975).
5. Matt Zoller Seitz, "Hot Seat: David Simon Explains 'Treme,'" *Salon*, July 4, 2011 www.salon.com/2011/07/04/treme_season_2_david_simon_interview/.
6. Steven Bochco, Terry Louise Fisher, and Jacob Epstein, "The House of the Rising Flan." *L.A. Law*, 1986, 01:03.
7. Diane Haithman, "Bochco on His Own: Veteran Producer's New Firm Bows With 'Doogie Howser,'" *Los Angeles Times*, September 14, 1989, http://articles.latimes.com/1989-09-14/entertainment/ca-444_1_bochco-created-doogie-howser.
8. Grimes, "A Feast of Wit?"
9. Ibid.

Chapter 4

1. Hercules, "Community!! Awake!! Grimm!! Chuck!! and a Surprise Apperance [*sic*] by Annie's Boobs!! Full NBC Comic Con Panels!!" *Ain't It Cool News,* July 25, 2011, www.aintitcool.com/node/50546.
2. Gina Wendkos, Stephen J. Cannell, and Frank Lupo, "Squeeze," *Wiseguy*, CBS 1989, 01:17.

Chapter 5

1. Dave, "Rick Moody Is Just Some Guy," PowellsBooks.Blog, October 10, 2006, www.powells.com/blog/interviews/rick-moody-is-just-some-guy-by-dave/.
2. Maureen Ryan, "Exclusive: 'Game of Thrones' Creator George R.R. Martin Addresses 'Lost' Feud and Finales," AOL TV, April 17, 2011, www.aoltv.com.2011/04/17/exclusive-game-of-thrones-creator-gerge-r-r-martin-lost/.
3. Ibid.
4. "The World's Most Popular TV Show Is . . . 'House'?" PopWatch, *Entertainment Weekly*, June 12, 2009, popwatch.ew.com/2009/06/12/house-worlds-most-popular-show/; and David Knox, "House Named as World's Most Watched Show," *TV Tonight*, October 3, 2011, tvtonight.com.au/2011/10/house-named-as-worlds-most-watched-show.html.

Chapter 6

1. Tom Fontana, "The Routine," *Oz*, HBO, 1997, 01:01.
2. Nick Hornby, "An Interview with David Simon," in *The Wire: Truth Be Told*, edited by Rafael Alvarez (New York: Grove, 2009), 384.
3. David Chase, "Pilot," *The Sopranos*, HBO, 1999, 01:01.
4. Ibid.

Chapter 7

1. Larry David, "The Pitch," *Seinfeld*, NBC, 1992, 04:03.
2. David Marc, *Demographic Vistas: Television in American Culture* (Philadelphia: University of Pennsylvania Press, 1996), 16–17.
3. Greg Metcalf, "If You Read It, I Wrote It: The Anonymous Career of Comic Book Writer Paul S. Newman," *Journal of Popular Culture* 29, no. 1 (Summer 1995): 147–162.
4. Sigmund Freud, *Civilization and Its Discontents* (New York: W.W. Norton & Company, 2005), 36–37.

Chapter 8

1. "The Making of Life's Too Short," promotional film, BBC/HBO, 2011.
2. Albert Camus, "The Myth of Sisyphus," in *The Myth of Sisyphus, and Other Essays* (New York: Alfred A. Knopf, 1955), 1–138.

Chapter 9

1. Tammy Oler, "The Vulture Transcript: Neil Gaiman on Comics, *Twilight*, Twitter Etiquette, Killing Batman, and Sharing Porn with His Son," *New York Magazine*, October 14, 2010, http://nymag.com/daily/entertainment/2010/10/the_vulture_transcript_neil_ga.html.
2. Adam Mehring, "Exclusive: Interview with Rodrigo Garcia, Writer and Director of MD Film Festival Closer 'Mother and Child,'" *The Dagger*, June 12, 2010, http://www.daggerpress.com/2010/06/08/exclusive-interview-with-rodrigo-garcia-writer-and-director-of-md-film-festival-closer-mother-and-child/.

Chapter 10

1. Sally Fitzgerald, ed., *The Habit of Being: Letters of Flannery O'Connor* (New York: Farrar, Straus and Giroux, 1988), 275.

2. Mark Singer, "The Misfit," *New Yorker*, February 14, 2005, 192–205.
3. David Milch, lecture to "Religion, Media and Hollywood" class, COMM 426 USC, January 24, 2008. Full video is available on theideaofthewriter.blogspot.com and clips are available on YouTube.
4. Ibid.
5. John Beluso and David Milch, "The Trial of Jack McCall," *Deadwood*, 2004, 01:05.
6. Milch, lecture to Religion, Media and Hollywood class.
7. Singer, "The Misfit."
8. William James, *Pragmatism, a New Name for Some Old Ways of Thinking: Popular Lectures on Philosophy* (New York: Longman's Green & Co., 1909), 290–291.
9. Ibid., 298.
10. David Milch, lecture to Religion, Media and Hollywood class.
11. Ibid.
12. David Milch, "His Visit: Day One," *John from Cincinnati*, 2007, 01:01.
13. Ibid.

Chapter 11

1. Dennis Potter, *The Singing Detective*, (New York: Vintage, 1988), 140.
2. For example: Jacob Weisberg, "*The Wire* on Fire: Analyzing the Best Show on Television," *Slate,* September 13, 2006, http://www.slate.com/articles/news_and_politics/the_big_idea/2006/09/the_wire_on_fire.html; Tim Goodman, "Decades in Review: The 10 Best TV Shows," *San Francisco Gate*, January 1, 2010, http://articles.sfgate.com/2010–01–01/entertainment/17466305_1_best-tv-shows-new-top-series-last-decade-2010; Mark Bowden, "The Angriest Man in Television: How David Simon's Disappointment with the Industry That Let Him Down Made *The Wire* the Greatest Show on Television—and Why His Searing Vision Shouldn't Be Confused with Reality," *The Atlantic*, February 2008, http://www.theatlantic.com/magazine/archive/2008/01/the-angriest-man-in-television/6581/; Jon Wilde, "*The Wire* Is Unmissable Television," *Guardian*, July 20, 2007, http://www.guardian.co.uk/culture/tvandradioblog/2007/jul/21/thewireisunmissabletelevis; "*The Wire*: Arguably the Greatest Television Programme Ever Made," *Telegraph*, April 2, 2009, http://www.telegraph.co.uk/news/uknews/5095500/The-Wire-arguably-the-greatest-television-programme-ever-made.html.

3. Matt Zoller Seitz, "Hot Seat: David Simon Explains 'Treme,'" *Salon*, July 4, 2011, www.salon.com/2011/07/04/treme_season_2_david_simon_interview/.
4. David Simon, "Introduction," in *The Wire: Truth Be Told*, edited by Rafael Alvarez (New York: Grove, 2009), 12–13.
5. Nick Hornby, "An Interview with David Simon," in *The Wire: Truth Be Told*, edited by Rafael Alvarez (New York: Grove, 2009), 383.
6. David Simon, "Mission Accomplished," *The Wire*, 2004, 03:12.
7. Petri Liukkonen and Ari Personen, "Maj Sjowall," *Books and Writer*, http://kirjasto.sci.fi/sjowall.htm.
8. Nick Hornby, "An Interview with David Simon," 384.
9. Dennis Lehane and David Simon, "Dead Soldiers," *The Wire*, 2004, 03:12.

Chapter 12

1. Kelly Faircloth, "Neil Gaiman Reveals the Future of *American Gods*," io9, June 24, 2011, http://io9.com/5815088/neil-gaiman-talks-the-future-of-american-gods?tag=neilgaiman.
2. Charlotte Higgins, "Neil Gaiman to adapt novel American Gods for HBO," August 22, 2011, *Guardian*, http://www.guardian.co.uk/books/2011/aug/22/neil-gaiman-american-gods hbo.
3. George R.R. Martin, George R.R. Martin in Conversation with Dan Anthony, Authors@Google, July 28, 2011, 24–29 minutes.
4. Vanessa Thorpe, "Salman Rushdie Says TV Dramas Comparable to Novels," *Observer*, June 11, 2011, sec. C7.
5. Todd VanDerWerff, "Dan Harmon Walks Us through Community's Second Season (part 3 of 4)," The A.V. Club, June 9, 2011, www.avclub.com/articles/dan-harmon-walks-us-through-communitys-second-seas,57252/.

Bibliography

Alvarez, Rafael. *The Wire: Truth Be Told.* New York: Grove Press, 2009.

Camus, Albert. "The Myth of Sisyphus." *The Myth of Sisyphus, and Other Essays*, 1–138. New York: Alfred A. Knopf, 1955.

Cooke, Lez. *British Television Drama: A History*. London: British Film Institute, 2003.

Douglas, Pamela. *Writing the TV Drama Series: How to Succeed as a Professional Writer in TV*. Studio City, CA: Michael Wise Productions, 2007.

Faircloth, Kelly. "Neil Gaiman Reveals the Future of *American Gods*." *io9*, June 24, 2011. http://io9.com/5815088/neil-gaiman-talks-the-future-of-american-gods?tag=neilgaiman.

Grimes, William. "A Feast of Wit? Not Quite. Bile Flows." *New York Times*, January 18, 1992, Sec. 1.13.

James, William. *Pragmatism, a New Name for Some Old Ways of Thinking: Popular Lectures on Philosophy*. New York: Longman's Green & Co., 1909.

Marc, David. *Demographic Vistas: Television in American Culture*. Philadelphia: University of Pennsylvania Press, 1996.

Martin, Brett. *The Sopranos: The Book*. New York: Time Home Entertainment, 2007,

Martin, George R. R., in Conversation with Dan Anthony, *Authors@Google*, July 28, 2011. http://www.youtube.com/watch?v=QTTW8M_etko.

McCabe, Janet and Kim Akass, eds. *Quality TV: Contemporary American Television and Beyond*. London: I. B. Tauris, 2007.

Metcalf, Greg. "If You Read It, I Wrote It: The Anonymous Career of Comic Book Writer Paul S. Newman." *Journal of Popular Culture* 29, no. 1 (Summer 1995): 147–162.

Milch, David. "Religion, Media and Hollywood" Class, COMM 426 USC, video. January 24, 2008, available on theideaofthewriter.blogspot.com.

The Museum of Television & Radio Seminar Series. *The Television of Dennis Potter: Seminar No. 2 of 4: Potter and Bochco: A Cross-Cultural Dialogue*. New York: Paley Center for Media, January 15, 1992.

Newcomb, Horace. *The Producer's Medium: Conversations with Creators of American TV.* New York: Oxford University Press, 1985.

Newcomb, Horace. *Television: The Critical View.* New York: Oxford University Press, 2006.

Ryan, Maureen. "Exclusive: 'Game of Thrones' Creator George R.R. Martin Addresses 'Lost' Feud and Finales." *AOL TV*, April 17, 2011. www.aoltv.com.2011/04/17/exclusive-game-of-thrones-creator-gerge-r-r-martin-lost/

Seitz, Matt Zoller. "Hot Seat: David Simon Explains 'Treme.'" *Salon*, July 4, 2011. www.salon.com/2011/07/04/treme_season_2_david_simon_interview/.

Singer, Mark. "The Misfit." *New Yorker*, February 14, 2005, 192–205.

VanDerWerff, Todd. "Dan Harmon Walks Us through Community's Second Season (Part 3 of 4)." *The A.V. Club*, June 9, 2011. www.avclub.com/articles/dan-harmon-walks-us-through-communitys-second-seas,57252/.

Index

Adam-12 (1968–1975), 45, 180
Adlard, Charlie, 159
The Adventures of Ozzie and Harriet (1952–1966), 125
Aguirre-Sacas, Roberto, 157, 210, 211
Aim High (2011–), 212
Alf (1986–1990), 57, 180
All in the Family (1968–1979), 94, 125, 168
All the King's Men (novel), 101, 170, 171, 190
Ally McBeal (1997–2002), 47, 57, 60, 81, 100
Ambiguity: *Breaking Bad*, 110, *Gilmore Girls*, disambiguation, 85; *The Good Wife*, 118–20; *John from Cincinnati*, 181–84; longform television, 44–45; *Mad Men*, 116–17; *Murder One*, 50–52; music, 26, 58–9; *The Singing Detective*, 26; *Treme*, 199–203; *Twin Peaks*, 69–70; *Veronica Mars*, 88
American Gods (2013–), 210–11
American Idol (2002–), 59, 127, 150
Amos & Andy (1951–1953), 125
Analyze This (film), 105
Anthology series: 64, 115, 155; *Cupid*, 86; disguised, 64–65; *Fringe*, 67; *House*, 90–91; *Millennium*, 66–67; *Six Feet Under*, 115; *Veronica Mars*, 88; *Wiseguy*, 73; *X-Files*, 65–67
Archie Bunker's Place (1979–1983), 125
Arrested Development (2003–2006, 2012), 131–32; Netflix and, 212
Attitudinal comedy, 133–34, 139–48; female-centered, 134, 140–44; male-centered, 143–48; soap opera, 140–43
Audience demographics, 176–77, 178–79
Auteur creator, 29, 83–85, 90
Autobiography: altered, 10, 29–30, 143–44; Bochco, Steven, 47–48; comedy, 149–53; Milch, 169, 180, 184
Awake (2012–), 213

Baretta (1975–1978), 64, 77
Barney Miller (1974–1982), 45, 145
Battlestar Galactica (2004–2009), 30; music, 32, 58–59, 211
BBC, 16–23, 25; middle-aged programming, 18; popular programming, 18; radio crossover, 18; short series, 20–21; social documentary 19–20; theater programming, 16–18. *See also* British television
The Beach Boys, 10
The Beatles, 10
Beck, Martin. *See* Wahloo, Per, and Maj Sjowall
Benny, Jack, 10
Ben 10 (2005–2007), 157
Be Tipul (2005 and 2008), 115
The Biederbecke Trilogy (1985, 1987, 1988), 135, 137
Big Brother (U.S.; 2000–), 149
The Big C (2010–), 139, 141
Big Love (2006–2011), 157
Birds of Prey (2002–2003), 160
Blackadder (1983–1989), 94, 135
Black Books (2000–2004), 135
Boardwalk Empire (2010–), 8, 58
Bochco, Steven, 16, 37–38, 47–48; "Television is not art medium," 55
Bonanza (1959–1973), 12, 63

Bones (2005–), 84, 104
Borges, Jorge Luis, 179
Boston Legal (2004–2008), 47, 81
Breaking Bad (2008–2012), 29, 31, 93, 109–14, 141, 172; change, 118, 120, 138; evil, 111–12; graphic novel, 155, 162, 164; hallucination, expected, 30; Shakespearean, 110
Brecht, Bertolt, 17
Brimstone & Treacle (1976/1987), 51
British television: comedy for older audience, 18; theater, 16
Buffalo Bill (1983–1984), 125
Buffy the Vampire Slayer (1997–2003), 29, 30, 89, 99–100, 157; continued as comic book, 165; musical episode, 33, 59, 60
Burn Notice (2007–), 81
Burns, Ed, 189

Californication (2007–), 142, 143, 144, 198; self-referentiality, 153
Calvin & Hobbes (comic strip), 183
Camus, Albert, 98, 140, 193
Canterbury's Law (2008), 93
Carnivale (2003–2005), 155, 162–64; comic book continuation, 165–66; magical realism, 163
Carter, Chris, 65, 67
Chandler, Raymond, 27, 56, 87
Change: *Breaking Bad*, 110, 111–14, *Friends*, 123, *The Good Wife*, 119–20; lack of change as tragedy in *The Sopranos*, 103–4; *NYPD Blue*, 168–69; *Sex and the City*, 141; *Slings & Arrows*, 138–39; *The Wire*, 198–99
Chase, David, 29, 71, 102–3, 105, 107
Cheers (1982–1993), 77, 130, 142
Chekov, Anton, 9, 106
Chicago Hope (1994–2000), 59
Chico and the Man (1974–1978), 151
Childrens Hospital (2008–), 211
Chinatown (film), 102
Chronicle of a Summer (film), 192
City of Angels (play), 56
The Closer (2005–2011), 13, 23–24
Close-up vs. wide shot: ideology of, 42
Columbo (1971–2003), 38, 77–78
Comcast, 212
Comic book, 121–22; continuation of canceled series, 165–66; television adaptations, 155; visual storytelling, 160–262; writing for television, 156–58. *See also* Graphic novels
Community: audience as, 127, 128; as hero in *Deadwood*, 170, 172–73, 175–76; heroine estranged from, 140; inadvertent in *Luck*, 184; instead of hero in *Treme* and *The Wire*, 58, 85, 196, 199, 203; shaping identity, 42; soap opera, 133; spiritual entity in *John from Cincinnati*, 177, 179–80. *See also* Internet, community
Community (2009–), 30, 34, 214
Concept album, 10, 15, 28
The Confession (2011), 212
Conway, Gerry, 157
Cop Rock (1990), 33, 47, 54–57; cancellation of, 60
Corinthians, the Book of, 172–73
The Cosby Show (1984–1992), 77
Crime Story (1986–1988), 185
Cupid (1998), 86
Curb Your Enthusiasm (2000–), 149–51; likeability, 126; as reality, 12, 30, 146

Dallas (1978–1991), 39
Damages (2007–), 49, 53–54, 61
Dancing with the Stars (2005–), 150
Darabondt, Frank, 159, 210
Darkness on the Edge of Town (record album), 10
The Days and Nights of Molly Dodd (1987–1991), 133–34; influence, 139–41, 146
Deadwood (2004–2006), 29, 78, 100, 165, 168, 169–76; community, 172–73; language 173–75; roots of, 169–71
Degeneres, Ellen, 151–52
Denise Richards: It's Complicated (2008–2009), 151
DeVries, Peter, 175

Dexter (2006–), 78, 79, 172, 209
Dickens, Charles, 8, 69, 189, 195, 199, 210
The Dick van Dyke Show (1961–1966), 57, 60
Dobbs, Michael, 212
Documentary influence: cinema verite, 192; direct cinema, 41, 131, 145; false, 41, 98, 99, 124, 132, 192; social, 19–20, 100, 148–49, 189. *See also* "Reality" television
Doogie Howser, M.D. (1989–1993), 47–48, 61; autobiographical 29, 37
Dr. Horrible's Sing-Along Blog (2008), 211
Dragnet (1951–1959), 12, 63, 94
Dramedy, 45–47, 48, 81–82, 112–13, 206; attitudinal comedy as, 133–34; comic function in, 45–47, 82–83, 147–48; comic structure without humor, 110–11; lack of humor, rarity, 46, 99–100; not exactly, 91, 105, 138; soap opera, 139–40; through syndication, 129–31
Dream On (1990–1996), 33–34
The Drew Carey Show (1995–2004), 59
DVD novel, 9, 137–39, 158, 177, 205, 211; distinction from successful television series, 56, 79, 90, 185, 188, 199; film as, 72, 206; graphic novel, 156, 159; greatest achievement, 199; permission, 214
DVD set, 6; bootleg, 56; extra features, 158, 180; film implications, 205–10; longform clarity, 94–96, 104, 110, 114, 118, 158, 188, 194, 203; music licensing, 75; narrative implications, 6–7, 53, 60, 67, 127, 156, 161; rescue series 52, 61, 79, 89, 90, 176, 188; serving size, 7

Edlund, Ben, 157
Ellen (1994–1998), 151–52
Ellison, Harlan, 182
Endings, 85–86
Enlightened (2011–), 13, 139, 141–42
Entourage (2004–2011), 143–44, 198
Ephemerality of television episodes, 2–5, 6, 9, 52–53; effect on musical television, 60; serial narrative, 8, 156
Episodes, free-standing: quality judged by single episode, 44, 95, 118, 194; values judged by single episode, 43–44, writing challenges, 2, 121–24
ER (1994–2009), 52, 90
Erendira (film), 179
Ex Machina (graphic novel), 157
Expectations, undermined, 194–97, 200–203

Fables (graphic novel), 160
The Fall and Rise of Reginald Perrin (1976–1979), 22, 33, 135–36, 137
Fallon, Jimmy, 127
Family Ties (1982–1989), 77
Father Knows Best (1954–1960), 2, 125
Fawlty Towers (1975–1979), 94, 135
The FBI (1965–1974), 94
Fearless (record album), 10
Ferguson, Craig, 127
Fifth Business (novel), 139, 197
Film on television, 3, 4, 6, 72
Fincher, David, 212
Firefly (2002), 6, 29; continued in comics, 165
Flipping Out (2007–), 151
Forbrydelsen (2007, 2009, 2012). See *The Killing*
Franklin & Bash (2010–), 81, 82
Friends (1994–2004), 130, 142; as soap opera in syndication 129; writing challenge of no change, 123
Fringe (2008–), 30, 67
Frost, Mark, 68–69
The Fugitive (1963–1967), 64–66

Gaiman, Neil, 136, 155, 160, 205, 210–11
Game of Thrones (2011–), 85, 159, 202, 211
Garcia, Rodrigo, 163
Genres: combined, 34, 81–83, 87, 90–91, 93, 137; juxtaposed, 27; revived with

soap opera, 11–13, 39–40, 107; weakened with soap opera, British, 18
The George Burns and Gracie Allen Show (1950–1958), 134
Gerber, Steve, 157
Gervais, Ricky, 20, 135, 148
GI Joe (1985–1986), 157
Gilmore Girls (2000–2007), 35, 81, 83–86, 89, 90; auteur voice, 84–85; class, 87; genres, 91; language, 94; music, 58; network interference, 84–86, 89, 90
The Girl Who Circumnavigated Fairyland in a Ship of Her Own Making (novel), 10
Glee (2009–), 33, 55, 59, 157
Goddard, Drew, 157
The Golden Girls (1985–1992), 77
The Good Wife (2009–), 7, 13, 118–20
Gossip Girl (2007–), 87
Graphic novels, 8–9, 10, 156; adapted to television 158–60; DVD novel, 156; visual storytelling, 160–62, 162–65
The Green Mile (novel/film), 210
Grierson, Edward, 19
Grillo-Marxuach, Javier, 157
Grimm (2012–), 63
The Guild (2007–), 211

Hammett, Dashiell, 8
Hannibal (2012–), 79, 209
Happy Days (1974–1984), 130
Harmon, Dan, 214
Harry's Law (2011–), 81
HBO, 3, 4, 104, 169–71, 212; demographics, target, 165, 176–78, 179; "quality" television, 25, 97–98, 193, 211
Heaven's Gate (film), 54
Heroes (2006–2010), 152, 155, 157–58, 162; "wide" narrative, 210, 211–12
High School Musical (2006), 59
Hill, Joe, 160
Hill Street Blues (1981–1987), 16, 34, 37, 38–46; audience, 38, 60; *Barney Miller* and, 145; comedy, use of, 45–46; *Homicide* and, 192; influence, 72, 81–82, 99, 145, 168–69, 172; musical adaptation, 55–56; realism, 41–43; soap opera, 11–12, 39–41, 69, 131; thematic variations, 44–45; *The Wire* and, 198
A History of Violence (graphic novel), 164
The Hitchhiker's Guide to the Galaxy (1981), 135, 136
Homicide: Life on the Streets (1993–1999), 188, 190, 193
The Honeymooners (1953–1956), 125, 146
House (2004–2012), 7, 31; *Blackadder* and, 94; "reality" game show, 92; season arc vs. episodes, 94–96; self-referentiality, 91; Sherlock Holmes and, 91; *The Singing Detective* and, 90–96; soap opera, 90–91; unlikable hero, 93–94
House of Cards (2012–2013), 212
Huggins, Roy, 64
Hung (2009–), 143

I Love Lucy (1951–1957), 4, 33, 57, 60, 125, 151; story structure, 103, 121–23
Inland Empire (2006), 71–72, 200
In Plain Sight (2008–2012), 81
Internet: community, 85, 86, 182, 202; serial publishing, 10; television viewing, 5, 6, 8, 52, 127, 205; webisodes, 210, 211–12; "wide" narrative, 77, 152, 158, 194
In Treatment (2008–2011), 46, 115–16
Invitation to Love (*Twin Peaks*), 69–70
Ironic sincerity, 132–33

James, Frank, and Jesse, 175
James, William, 175, 177–78; "Fifth Lecture on Pragmatism," 177–78; and Henry, 175; *Varieties of Religious Experience*, 177
The Job (2001–2002), 145–47
John Dies at the End (novel), 10
John from Cincinnati, 6, 176–84, 185; casting, 183–84; magical realism, 179–80; music, 184
Johns, Geoff, 157
Judging Amy (1995–2005), 81

Keeping Up with the Kardashians (2007–), 151
Kelley, David E., 47, 53, 60, 167
Kendra (2009–), 151
The Killing (2011–), 46; comic book style, 155, 162; *Forbrydelsen*, 164; opacity, 164; seriousness, 116, 118
Kindle, 205
King, Stephen, 157; adaptation of works by, 209–10
Kirkman, Robert, 159
Knauf, Daniel, 163
Knott's Landing (1979–1993), 39
Kolchack, the Night Stalker (1974–1975), 65
Kopple, Barbara, 192
Kring, Tom, 157
Kubrick, Stanley, 187–88
Kuleshov effect, 130

L.A. Law (1986–1994), 37, 42, 61, 81; comedy, cynical, 46–47
Landau, John, 10
The Larry Sanders Show, 29
Laughtracks, 126–28, 133–34, 140, 145; absence of, 127–28; oceanic effect, 127
Law & Order (1990–2010), 53, 157
The League of Gentlemen (1999–2002), 135
Leave It to Beaver (1957–1963), 48
Legend of Neil (2008–2010), 211
Lehane, David, 193
Lewis, R.W.B., 175
Lie to Me (2009–2011), 93
Life's Too Short (2011–), 8
Likeability, 131; killing of likeable characters, 201–3; likeability machine, sitcom as, 124–26
Like Water for Chocolate (film), 179
Lilyhammer (2011–), 212
The Limey (film), 174
Lindelof, Damon, 160
Locke & Key (graphic novel), 160
Loeb, Jeff, 157
Longform film on DVD, 71–72, 206–8
Longform television: blended genres, 81–84, 87, 91–92, 94; episodes balanced with long arc, 82, 88, 94–96, 105–6, 113–14; judged for series, not episode, 43–44; multiple stories, 39
Lost (2004–2010), 30, 86, 158, 183, 213; comics, 155, 157, 162; epilogue on DVD, 158; online presence, 182, 211–12; "wide" narrative, 152, 158, 210
Luck (2012), 184–85
Lynch, David, 68–72, 208

Macdonald, Ross, 87
Mad Men: Don Draper precedent, 78; as koan series, 116–17; male soap opera, 115–16, 116–18
Magical realism, 163, 179–80, 183
Magnum, P.I. (1980–1988), 45
Man Facing Southeast (film), 179
The Man from U.N.C.L.E., 12, 63; comics, 121
Mann, Michael, 184–85
Marc, David, 121, 123
Marquez, Gabriel Garcia, 163, 179
Martin, George R. R., 159, 211; on endings, 85–86
Mary Hartman, Mary Hartman (1976–1977), 132, 148
Mary Tyler Moore (1970–1977), 126, 130
*M*A*S*H* (1972–1983), 128, 130
Masterpiece Theatre (1971–), 22
McNeill, Carla Speed, 10
Merchant, Stephen, 20, 148
Miami Vice (1984–1990), existentialism and, 98–99, 32–33; multi-episode arc, 63; music, 59; music rights, 75
The Middleman (comic book), 157
Milch, David, 167–85; auteur control, 179; community, 175, 179; *Hill Street Blues*, 38; William James, 177–78; language, 167, 169, 173–75, 181; R.W.B. Lewis, 175; literary goals, 170–76, 177; Monism, rejection of, 177–79; passive-aggressive creation, 176–77; Roman series, 170–71;

symbols ("illusions agreed upon"), 171, 182–83; Robert Penn Warren ("pain of past"/"future tense of joy"), 168, 170, 175, 176
Millennium (1996–1999), 66–67, 73
Miller, Frank, 161
Mills, David, 191
Miniseries, 4, 22, 210
Monty Python's Flying Circus (1969–1974), 17, 18
Moonlighting (1985–1989), 81, 152–53
Moore, Alan, 28, 161, 207
Moore, Tony, 159
Mork & Mindy (1978–1982), 153
Mr. Roger's Neighborhood (1968–2001), 127
Murder, She Wrote (1984–1996), 77
Murder One (1995–1997), 29, 48–53, 61; *Brimstone & Treacle* and, 51; influence on *Damages*, 53–54; OJ Simpson trial and, 49–50, 52; soap opera influence, 50
Musicals, Broadway: *City of Angels* as *The Singing Detective* adaptation, 56; *Cop Rock*, 55; *Dr. Horrible's Singalong Blog*, 211; from film 56; *Hill Street Blues* near-adaptation, 43; *The Producers*, 151; special episodes as, 59–60; *Spiderman: Turn Off the Dark*, 157
Music on television, 15, 32–35, 54–60; familiarity, need for, 57; *John from Cincinnati*, 184; licensing fees, prohibitive, 58; *The Singing Detective*, 25–27; soap opera music, 50, 69; *Treme*, 199; *Wiseguy*, 74–75
My Favorite Martian (1963–1966), 180
My Mother the Car (1965–1966), 180
Mystery (Masterpiece Mystery) (1980–), 22
"The Myth of Sisyphus" (essay), 98, 140
Mythos. *See* Anthology

"N." (short story), 210
Narrowcast vs. broadcast, 3, 67
Natural Born Killers (film), 34
Neorealism, 41–42, 200–201; applied to audio, 42, 192, 200
Netflix, 6, 212
Neverwhere (1996), 210
Newman, Paul S., 121–22
Newman, Randy, 55, 60
The Next People (2013), 212
Nobbs, David, 135–36
Nonfiction novel, 189
Norris, Frank, 189
Northern Exposure (1990–1995), 58
Novelists writing television, 22–23, 86–87, 213
Novels, adapted to television, 21–22, 158–59, 208–11, 212
Nurse Jackie (2009–), 139, 141
NYPD Blue (1993–2005), 168–69

O'Connor, Flannery, 167
The Office (UK; 2001–2003), 135; social documentary, 20, 148–49
The Office (U.S.; 2005–), 149
Oliver! (film), 188
Oliver's Travels, 137
O'Malley, Mike: Tommy Carcetti and, 189–90, Huey Long/Willie Stark and, 190
Once Upon a Time (2011–), 160
Online. *See* Internet
Oversimplification, 2, 51
Oz (1997–2003), 33, 97, 98–100; male soap opera, 98–99; sincerity 99–100

Pease, Otto, 175
Pelecanos, George, 193
Permission, creative, 28–29, 214
Pet Sounds (record album), 10, 15, 28
Peyton Place (1957), 69
Plater, Alan, 16, 22, 137
Police Squad (1982), 162
Potter, Dennis, 25, 28; influence on *Murder One*, 51
The Practice (1997–2004), 53
Pragmatism, 177–78
Prescription: Passion (*House*), 91
Price, Richard, 193
Prime Suspect (UK; 1991–2006), 23–24; *The Closer* and, 23–24; *Prime Suspect* (U.S.) and, 24
Prinze, Freddie 43

Prisoner (a.k.a *Prisoner: Cellblock H*) (1979–1986), 99
The Prisoner (1967), 17–18
The Producers (film/play), 151
Profit (1996), 78–79
Prometheus, 182–83
The Pussycat Dolls Present: The Search for the Next Doll (2007), 89

Quality television, 18, 97, 122; conventions of, 200–201

Real Housewives franchise (2006–), 149, 151
Realism: effects of OJ Simpson trial, 48–50; effects of social documentary, 19–20, 148–49; *Hill Street Blues*, 43; musical television, 59; *Oz*, 99
Reality, altered, 26, 30–32, 33–34; dreams, 70–71, 115, 163, 213; halluci nation, 30–33, 60, 70, 95, 213
"Reality" television, 12–13, 89; comedy and, 151–53; *Curb Your Enthusiasm* and, 149–51; *House* and, 92; influence on television musicals, 59, 61; *Moonlighting* and, 152–43; *The Office* (US) and, 20; "reality" game shows, 13
Remington Steele (1982–1987), 153
Rescue Me (2004–2011), 30, 31, 145–48, 198
Revenge (2011–), 209
Revolver (record album), 10
Risky Business (film), 111, 184
Rock Follies: influence 33; music 26
Roseanne (1988–1997), autobiography, 29, 31, 151; class, 129; family, 142; syndication, 129
Rubicon, 116, 118; opaque narrative, 155
Rushdie, Salman, 213

Sailor Twain, or The Mermaid in the Hudson (graphic novel), 10
The Sandman (graphic novel), 160, 210
Saving Grace (2007–2010), 147–48
Schizopolis (film), 174
Screwball comedy, 83–84, 86
Scrubs (2001–2010), 28, 30, 33, 34, 59
Secret Girlfriend (2009–), 211
Seinfeld (1990–1998), 52, 134, 142; autobiography, 30, 151–52; *Curb Your Enthusiasm*, 150; ending, *The Sopranos* and, 104, 122–23; syndication, 129; unlikable characters, 124, 126
Self-referentiality. See *Twin Peaks*; Soap opera
Serial narrative, compiled, 8–10
Sex and the City (1998–2004), 141
Sexual abuse: depiction of, *Homicide* vs. *NYPD Blue*, 191
Sgt Pepper's Lonely Hearts Club Band (record album, 1967), 10, 15
Shakespeare: Great Man tragedy, 108–9, 112–13; *Hamlet*, 114, 138; *King Lear*, 138; *Macbeth*, 54, 99, 100, 111–12, 138; *The Merchant of Venice*, 44; *Taming of the Shrew* parody, 152; Shakespearean television, 100–101, 106, 173
Shakur, Tupac, 10
Shameless (UK; 2004–), 135
Shark (2006–2008), 93
The Shield (2002–2008), 107–9, Shakespearean character, 108–9
Short Series, 20–21, 209
Siegel, Mark, 10
Simon, David, 9, 186–203
The Simpsons (1989–), 123–24, 126
Sinatra, Frank, 10
Sinclair, Upton, 189
The Singing Detective (1986), 15–16, 24–28, 167; alternate reality, 30–32; autobiography 25–26; influence on U.S. television, 28–35, 115; influence on auteur creators, 29; influence on music, 32–34; influence on Steven Bochco 37, 55, 58; *Ally McBeal* and, 60; *Awake* and, 213–14, *The Job* and, 169; *NYPD Blue* and, 208; *The Sopranos* and, 30, 146; *Twin Peaks* and, 70; *Watchmen* and, 213; reaction 17; self-referentiality, 25–26; unlikable hero, 24–25, 29
Single story season: *Damages*, 53–54; *Heroes*, 157–58; *Murder One*, 49–52;

Terriers, 89–90; *Twin Peaks*, 68; *Veronica Mars*, 87–88
Situation comedy, 121–33; as likeability machine, 124–26; British, 18, 20, 135–37; change, lack of, 121–24, 126, 141; family ideal, 124–25; syndication and free-standing episodes, 121, 126; soap opera, 131–33; soapiness of syndication, 128–31, 141, 149
Six Feet Under (2001–2005), 98, 113, 115, 117
Slings & Arrows (2003–2006), 137–39
Smash (2012), 33, 61
Snyder, Zach, 161
Soap (1977–1981), 131
Soap opera, 2, 50, 83; blended with genre, 82, 88, 90, 171; British, 18, 20; definition, 11; male-centered, 11–12, 40, 82, 106–7, 114–16, 197–99; *Oz* and, 98–99; in prime time, 39; and "reality" programming, 13, 59; and Steven Bochco, 61; *Hill Street Blues*, 34, 39–40; *LA Law*, 46; primetime, 39; *Profit*, 78; self-referentiality, 69, 91, 131–33; *Twin Peaks*, 68–70, 94; unexpressed emotion, 116–20; women in male-centered, 53, 118–20, 198. *See also* Attitudinal comedy, soap opera; Situation comedy, soap opera; Situation comedy, soapiness of syndication
Soderbergh, Steven, 174
Songs for Young Lovers (record album), 10
Sons of Anarchy (2008–), 114
The Sopranos (1999–2007), 28, 29, 35, 98, 100–107, 212, 213; altered reality, 30; as comedy 104–7; change, lack of, 103–4; Chekov, 106; closure 102–4; ending, 102–4, 124; morality, conflicted, 79; pilot, 71; Shakespeare, 100–101, 195; tragedy of average man, 101, 102; on VHS, 4
Spaced (1999–2001), 135
Spacey, Kevin, 72, 76, 212
Spider-man (comic book), 157
Spider-man: Turn Off the Dark (play), 157
Spin City (1996–2002), 84
Spoilers, 127
Spooks (MI-5) (2002–2011), 201
Sports Night (1998–2000), 81, 128
Springsteen, Bruce, 10
The Stand (novel), 157; film adaptation, 210
Starman (film), 181
The Story of a Crime (novels). *See* Wahloo, Per, and Maj Sjowall
Studio 60 on the Sunset Strip (2006–2007), 82
Suits (2011–), 81, 82
Supernatural (2005–), 58
Survivor (2000–), 13, 149

Talisman (graphic novel), 10
Taxi (1978–1983), 125, 130–31
Telephone series, 5, 210
Television: death of, 37, 205; definitions of, 214, episodic writing, 40–41; network, 1, 3–4, 6, 79, 104; writing for network, 3–4, 117, 178; writing long-form, 41–42
Terriers (2010), 89–90
Theater on television, 16–18
30 Rock (2006–), 34
Thomas, Rob, 86, 89, 166
300 (comic book/film), 161
Thug Life: Volume 1 (record album), 10
The Tick (1994, 2001), 157
Till Death We Do Part (1965–1975), 94
Tivo, 5, 52, 67, 127
Tommy (record album), 10
Tragedy: Great Man and Average Man, 112–13; Greek tragedy, 196; lack of change as in *The Sopranos*, 103–4; lack of change as in *The Wire*, 198–99; post–World War II, 101–2
Transformation. *See* Change
Transformers (1984–1987), 157
Treme (2010–), 199–203; community, 173; dread, 200–203; DVD novel, 9, 188; long arc, 44; music, 58, 199, 202; time passage, 199

True Blood (2008–), 57, 209 *24* (2001–2010), 6, 49, 54, 163
Twin Peaks (1990–1991), 38, 68–71, 132; self-referentiality, 69–70; soap opera, 68–69; surrealism, 68
Two and a Half Men (2003–), 144, 153
2001, A Space Odyssey (film), 188

Under the Dome (2013), 209
The United States of Tara (2009–2011), 13, 30–31, 35, 139, 142, 148
Unlikability of hero, 29, 78–79, 93, 94, 107–8, 124–26, 168–69, 176
The Untouchables (1959–1963), 12
UPN, 87, 88, 89

Valente, Catherynne M., 10
The Vampire Diaries (2009–), 209
Vaughn, Brian K., 157
Veronica Mars (2004–2007), 35, 81, 86–90, 91; balance of episodes and longform story, 88; class awareness, 87; comic book continuation, failed, 165–66; network interference, 89; novelistic, 89; voice, 94, 174
VHS videotape, 1, distribution of premium cable shows, 4; shows released on, 4–5
Video games, 136, 158, 161, 206
Viewing, need for attention, 117–20, 162, 164–65, 181–83, 192, 193–94
Villains, attractive, 72, 73–77, 78–79. *See also* Unlikability of heroes

Wahloo, Per, and Maj Sjowall, 195
The Walking Dead (2010–), 156, 159–60, 162, 164, 211
Warren, Robert Penn, 101; *All the King's Men*, 101, 170, 171, 190; evaluation of Milch, 168, 174; influence on Milch, 168, 170, 171, 175, 181
Watchmen (graphic novel/film), 28, 161, 207
The WB, 84, 88–89
Webber, Andrew Lloyd, 184
Web Therapy (2008–), 211
Weeds (2005–), 13, 142–43, 144
Wein, Len, 157
Welcome Back, Kotter (1975–1979), 151
Whedon, Joss, 29, 165, 157, 211
White Collar (2009–), 82
"Wide" narrative, 152, 158, 210, 212
Willingham, Bill, 160
Winfrey, Oprah, 13
The Wire (2002–2008), 8, 29, 100, 187–88, 193–99; *Deadwood* and, 173, 196; Dickens, 8, 189, Greek tragedy, 100, 173, 195–96; heroic ideal, rejection, 173; literary precedents, 195; male melodrama, 193, 196–99; urban sociology, 197–99; visual novel, 193
Wiseguy (1987–1990), 29, 72–77, 99; disguised anthology, 73; influence, 108; long arcs, 73; music, 32, 74–75; sympathetic villains, 72–77, 108
WKRP in Cincinnati (1978–1982), 33, 59, 75, 130, 184
Wolverine (comic book), 150
The Wonder Years (1988–1993), 48
Wong, David (Jason Pargin), 10

The X-Files (1993–2002), 57, 65–67, 72, 92; anthology, disguised, 65–67, 73, 77; Christian mythos, 66; the Smoking Man, 66, 72, 77

Y: The Last Man (graphic novel), 157
YouTube, 60, 75, 158, 212

Zucker Brothers, 162

About the Author

Greg Metcalf is an artist and a scholar who teaches film, television, literature, modern art history, cultural history, popular culture, American humor, and sculpture at the University of Maryland and the Maryland Institute College of Art when he is not making ritual sculptures or writing about the arts in culture.

He has a short attention span.